NEW DIMENSIONS IN AGRICULTURAL GEOGRAPHY

VOLUME SEVEN

DYNAMICS OF AGRICULTURAL DEVELOPMENT

Concept's International Series in Geography

Concept's International Series in Geography No. 4

NEW DIMENSIONS IN AGRICULTURAL GEOGRAPHY

VOLUME SEVEN

DYNAMICS OF AGRICULTURAL DEVELOPMENT

NOOR MOHAMMAD
Editor

CONCEPT PUBLISHING COMPANY, NEW DELHI 110059

ISBN 81-7022-400-4 (Set)
ISBN 81-7022-407-1 (Vol. 7)

First Published 1992

Published by
Ashok Kumar Mittal
Concept Publishing Company
A/15-16, Commercial Block, Mohan Garden
NEW DELHI 110059 (India)

Lasertypeset by
Microtech Advance Printing Systems (Pvt.) Ltd.
H-13, Bali Nagar,
NEW DELHI 110015

Printed at
Sunil Printers
CB 386, Ring Road, Naraina
NEW DELHI 110028

PREFACE

AGRICULTURE CONTINUES to be the key industry. It is not only the dominant form of employment but also forms the basis for a way of life for, perhaps, two-thirds of the world's households. Agriculture being the function of physical, socio-institutional and techno-economic factors which are dynamic in nature, keeps on changing with the basic objective of increasing production and generation of foodgrains surplus. The traditional system of farming is in the process of transformation to modern agriculture which itself is undergoing a set of equally and generally even in more rapid changes. So it has added a new dimension in agriculture which needs to be studied and analysed.

A project was, therefore, taken up to accomplish this task. Original research papers dealing with various facets of agriculture were invited from scholars of different discipline from all over the world. The response was very encouraging and a large number of original papers were received. Besides these, some papers which have classical and everlasting value have also been picked up from my earlier publication – *Perspectives in Agricultural Geography*. The papers have been classified according to their basic themes and have been put into following eight volumes so as to give a systematic picture of various themes. The articles dealing with concepts, theories and models have been kept first followed by those dealing with the same theme at macro, meso and micro levels respectively.

The First Volume – *Historical Dimensions of Agriculture* – containing 19 articles deals with the origin, diffusion and development of agriculture, research trends in agricultural geography, world agricultural systems and methods and techniques in agricultural geography.

The Second Volume – *The Ecology of Agricultural System* – deals with the selected parameters of physical environment specially the climatic elements which affect the agricultural landscape in spite of technological advancement in agriculture.

The Third Volume – *Socio-economic Dimensions of Agriculture* –

containing 19 articles deals with socio-economic environment which have bearings on agricultural systems. The first 8 articles deal with social parameters specially labour, caste, women etc., while the last 11 articles examine the impact of techno-economic factors specially the various facets of irrigation which is a basic agricultural input.

The Fourth Volume – *Landuse and Agricultural Planning* – contains 19 articles dealing with concepts, theories, models of landuse, land capability and suitability, wastelands and dynamics of landuse.

The Fifth Volume – *Spatial Dimensions of Agriculture* – includes articles on agricultural systems of specific areas such as Himalayas, Bastar, Uttara Khand etc. It also deals with dynamics of crop landuse, cropping pattern and its diversification, crop ecology and crop-combination analysis.

The Sixth Volume – *Human Perception and Technological Change in Agriculture* – contains only 15 articles and covers four basic themes. The first set of 3 articles deals with human perception and his consequential response, while the second set includes articles on correlates of technological change and adoption of agricultural innovations. The third set of articles deals with the assessment of socio-economic transformation in rural India. The last set deals with agriculture and human settlements.

The Seventh Volume – *Dynamics of Agricultural Development* – includes articles on concept of development and its various facets. Besides assessing the trends of levels of agricultural development, it also examines and evaluates its impact on environment and ecological parameters.

The Eighth Volume – *Anthropogenic Dimensions of Agriculture* – containing 17 articles deals with regional imbalances in agricultural productivity, methods of its measurement, and also its correlates and determinant. It also focuses on food availability and food security at regional and continental level.

The collection of papers in eight volumes of *New Dimensions in Agricultural Geography* is designed to encapsule the major themes and recent trends and concepts in the field of agricultural geography with particular reference to conceptual revolution in the discipline. All the volumes contain a brief introduction which gives a glimpse into the topics dealt with in each of them.

The variety of articles contributed by the renowned scholars of different disciplines has made the character of these volumes multi-disciplinary and has enhanced their value. These volumes have been designed to help not only the students of a particular discipline but also to general students who wish to keep abreast of the latest developments in the field of agricultural geography. From these points of view these volumes hold a promise of rich fare and in addition to being heuristic, will provide a mine of information for discerning scholars. It is hoped that this unique collection of research papers will not only be valuable to the researchers in the fields of agricultural geography, economics, sociology, history, and agricultural science, but will also provide guidelines to decision-makers and planners.

The completion of this project has been possible due to the constant inspiration and unstinting co-operation of a host of people too numerous to acknowledge individually. However, I owe a debt of gratitude to my mentor — Dr. Mohammad Shafi, Professor Emeritus, Department of Geography, A.M.U., Aligarh who gave me valuable comments and suggestions ever since the inception of this project. Prof. R. Ramachandran and my other colleagues who offered sincere comments on the title of volumes and the collection of articles deserve special thanks. I would also like to thank my father, wife and children who provided a favourable working atmosphere to me for successful completion of this project. Last but not the least, I would like to put a special word of appreciation, love and affection to Mohammad Imran Ali Khan who is a source of hope and inspiration.

Department of Geography NOOR MOHAMMAD
Delhi School of Economics
University of Delhi
Delhi-110007
January 26, 1992

PUBLISHER'S NOTE

When Concept's International Series in Geography No. 1 — *Perspectives in Agricultural Geography* edited by Dr. Noor Mohammad in five volumes was issued over a decade back, we expressed our apprehensions that 'Posterity will be the best judge of our efforts in this direction. The response to this set of volumes will provide the impetus for hastening the publication of further series that are already on the anvil'.

Perspectives in Agricultural Geography received such an overwhelming response that it became out of print soon after its publication. Its success provided the necessary encouragement to issue thereafter Concept's International Series in Geography No. 2 — *Perspectives in Urban Geography* edited by Dr. C.S. Yadav in 16 volumes and No. 3 — *Perspectives in Geomorphology* by Dr. H.S. Sharma in four volumes.

We are glad to offer now Concept's International Series in Geography No. 4 — *New Dimensions in Agricultural Geography* edited by Dr. Noor Mohammad in eight volumes. In view of the new emerging trends, concept, theories and models in the field of agricultural geography, the set presents a wide panoramic view of the latest advances in this field. Hope this set too would be equally welcomed by the students, researchers, scholars and planners in various disciplines.

February 11, 1992 NAURANG RAI

CONTENTS

LIST OF FIGURES

LIST OF TABLES

Map

LIST OF CONTRIBUTORS

Bhadrapur, V. S.
Department of Geography, Karnataka University, Dharwad

Casetti, Emilio
Department of Geography, The Ohio State University, Columbus, USA

Chand, Raghubir
Department of Geography, Kumaon University, Nainital

Ciriacy-Wantrup, S. V.
Department of Agricultural Economics, University of California, Berkeley, U.S.A.

De, N. K.
Department of Geography, Burdwan University, Burdwan

Joshi, Kailash Chandra
Giri Institute of Development Studies, Lucknow

Kaddi, V. B.
Department of Geography, Karnata Science College, Dharwad

Korishettar, S. F.
Department of Geography, Karnataka Science College, Dharwad

Krishan, Gopal
Department of Geography, Punjabi University, Chandigarh

Majeed, Abdul
Department of Geography, S.S. School, Jamia Millia Islamia, New Delhi

Mattimani, C. N.
Department of Geography, Kittel Arts College, Dharwad

Mohammad, Noor
Department of Geography, Delhi School of Economics, University of Delhi, Delhi

Murton, Brian J.
Department of Geography, University of Hawaii, Honolulu, Hawaii

Naregal, S. S.
Department of Geography, Karnataka University, Dharwad

Pannell, Clifton W.

Department of Geography, University of Georgia, Athens, Georgia, USA

Pathak, C. R.

Department of Architecture and Planning, Indian Institute of Technology, Kharagpur

Sarkar, A. K.

National Atlas & Thematic Mapping Organisation, Government of India, Calcutta

Singh, Sham Sher

Department of Architecture and Planning, Indian Institute of Technology, Kharagpur

Singh, Surender

Department of Geography, Shivaji College, University of Delhi, New Delhi

Thakur, Rameshwar

Department of Geography, Maithan Degree College, University of Ranchi, Maithan

Togarsi, G. V.

Department of Geography, Karnatak Science College, Dharwad

Veeck, Gregory

Department of Geography and Anthropology, Louisiana, State University, Batan Rough, USA

INTRODUCTION

AGRICULTURAL DEVELOPMENT is a multi-dimensional concept which mainly includes development in areal strength of cropped land, improvement in farming practices/systems, improved farm implements, irrigation system and irrigated area, high yielding/ improved varieties of seeds, chemical fertilisers, insecticides and pesticides, intensity of cropping and specialisation and commercialisation of agriculture. The development in all these parametres is ultimately reflected in agricultural productivity which itself is very complex and controversial concept (Mohammad, 1981). However, this volume opens up with **Krishan's** paper which has very precisely discussed the concept of agricultural development. Agricultural development, in a true sense, denotes the quality of agricultural system of a region in terms of productivity, diversification and commercialisation consistent with a desired state of agrarian relations and ecological balance. The level and rate of agricultural development may be distinguished as the former represents a picture prevailing at a particular point in time, while the latter stands for the progress achieved over a given period. The author has reviewed some selective but sufficiently representative literature on this theme and opines that all studies take a partial view in varying degree of agricultural development. Main emphasis is given on dimensions of productivity and its determinants. Finally he concludes that agricultural development is a fascinating field of study in geography.

Agricultural productivity is the function of physical, social and economic factors and is measured in various ratios depending upon the socio-economic condition of the area. Labour productivity in agriculture is most important. **Casetti's** paper is concerned with agricultural labour productivity and various facets of development. The author after giving an analytical review of Vendoor's Law (1945), Kristensen's (1974) effect studies on population pressure and productivity growth, has propound three hypotheses viz. the increase in labour productivity in agriculture is positively related with the rate of growth of agricultural output; the highest growth rate of agricultural labour productivity tend to occur in countries at intermediate level of development; and population density and rate of growth of population are positively related with growth rate of agricultural productivity. Cross-section data of 6 selected variables from a sample of 71 countries are used to test the hypothesis with regression equation technique. The analysis reveals that the rate of growth of agricultural labour productivity tends to be positively related with the growth rate of agricultural sector; tends to be a greater at intermediate levels of development; and is positively affected by population pressure when the rate of change of labour force is controlled for.

In the past two decades the economists were giving more emphasis to technological institutions or modern factors of production as compared to natural sources and economic institutions in the economic growth. This applies specially to renewable resources used in agriculture. Even in the current mathematical models of economic growth the role of institution is not explicitly considered. **Ciriacy-Wantrup** in his paper has focused on doctrine of the decreasing importance of the natural resources and the relatively insignificant and passive role of institutions in economic growth. In the continuously changing interplay of challenge and response between natural environment and human culture, both are equally significant and are a part of the same system — human ecology. He has also examined the role of economic institution in economic growth and according to him it operates on second level of three levels hierarchy of decision systems. On each level the structure and functioning, and performance of the decision system can be studied — both conceptually and observationally. To elaborate his points further he has studied in depth the role of water institution in northern India under the impact of Green Revolution — the rapid spread of new cereal technology based on varieties with greater capacity to combine with complimentary inputs, such as labour, fertiliser and water.

One of the dimensions emphasised in the growing literature on sustainable development in agriculture is the importance of local control and participation in development planning and specially for a small farmer oriented set of agricultural policies. This approach involves a research and development agenda for resources poor farmers which would be highly specific to local ecological conditions founded on a thorough knowledge of local agricultural practices, and participatory in style. **Murton** in the first section of his paper explains and examines with suitable examples the methods and techniques in which Indian farmers traditionally have managed soil, water, climate, slope and space. The second section reviews the local knowledge about the environment of agriculture in which categorisation of resource discrimination, coordination of agricultural activities and resource productivity are thoroughly discussed. The last section raises questions about the adequacies of a populist approach and suggests ways of resolving the difficulties involved in accomplishing such a task. The study concludes that sustainable agricultural development requires on the spot attention involving local knowledge, continuous monitoring and sometimes vigorous action. The policy-makers and researchers should listen and learn from farmers.

During the past decade there have been profound changes in the rural economy in China resulting to unprecedented prosperity, rise in income, and improvement in housing space and quality. The reforms which have been widely credited for these improvements are actually the result of a series of interrelated economic development policies designed to faster greater commercialisation, diversification and integration of rural areas within larger economic sphere. **Veeck** and **Pannell** in their paper have examined variations in foodgrain production and efficiency in Jiangsu in China. The study is based on primary data collected from sample households chosen from 8 counties of four prefectures. The data pertaining to household income and income source, household demographic information, output of wheat and rice etc., were collected for the year 1986-87. The study after examining other possibilities hypothesised that different types of households, once identified, are a more salient predictor of production efficiency for foodgrains. To identify the types of households and to classify the surveyed households into these types, a sorting routine employing Principal Component Analysis using 7 variables and Wards Minimum Variance Clustering Procedure is performed. The result indicates that three variables accounted for 72 per cent variance within the seven variables across the entire sample. The study, however, concludes that the rural reforms introduced in Jiangsu in 1979-80 have transformed many areas from primarily agricultural economics to mixed economies with greater percentages of employment in industry, service, transportation and construction. This shift has resulted in households which derive income from multiple sources. Households with industrial positions tend to have higher income and, at least for the households surveyed for this research, are less efficient producers of foodgrains. The higher incomes have led to programme designed to restore local party. The differences in production efficiency may, in part, be the impetus behind recent efforts to promote the "Specialised Household" programme. It appears that Chinese planners feel that sufficient time has elapsed to evaluate the 1979 reforms, and have already taken steps to develop appropriate adjustment policies. There is no doubt that the reforms have resulted in new types of places, neither urban nor rural, and that the employment opportunities in these areas are different as well. What is less certain is how both efficiency and total agricultural production will be affected, and who will be farming in the "Kotadesasi" of the future.

As a result of introduction of new programmes and policies, agriculture in India has developed, although unevenly, to a considerable extent. **Thakur** in his paper has presented the growth of agriculture in Bihar at district level. Three years moving average data for 1963 and 1973 for 17 major crops which occupy more than 80 per cent of the total cropped area have been analysed. In order to give clear and composite picture, the growth of agriculture has been assessed at four levels i.e. changing patterns of general landuse with special reference to fallow land and net area sown; changing patterns of crop landuse in kharif and rabi seasons; crop-wise growth in area and output; and overall growth at state level in monetary value. On the basis of the levels of growth in area, production and productivity, the author has demarcated the positive and negative areas of growth in each of these cases separately. Finally the author has correlated the growth in agriculture with four important agricultural innovations viz. irrigations, fertiliser, tractor and pumping sets, and has worked out correlation co-efficient. The study reveals that there is weak relationship between the agricultural growth and these variables.

The Green Revolution has brought tremendous development in agriculture but it is uneven at all levels. Considering that western Uttar Pradesh has witnessed a conspicuous upheaval in its agriculture during last two decades. **Chand** and **Joshi** have analysed the agro-economic trend specially with reference to changes in broad landuse categories, area, yield and production of crops, and have assessed the impact of irrigation, fertiliser and other technological inputs on production level. The three years moving averages for the years 1965-66 and 1985-86 for landuse, area, yield and production of almost all the major crops have been taken. In order to measure the variability patterns, trend line has been fitted with a regression line and correlation co-efficient is worked out and tested for significance by 't' test. The study reveals fluctuating trend in landuse with aggregate gain in net sown area, changing pattern of crop landuse with increasing emphasis on fine and more remunerative crops — wheat, rice and sugarcane. The production shows a positive and highly significant correlation co-efficient with area in all cases. The study concludes that the level of agricultural productivity has improved mainly due to manifold increase in modern technological inputs but fluctuation in production is attributed to inadequate rainfall and other environmental factors which lead us to conclude that weather elements have still a bearing on agriculture.

In an other article **Togarsi** et al. have studied the impact of regional disparities on the development of agriculture in a predominantly agricultural district of Bijapur. For the measurement of regional disparities 28 indicators grouped into 3 categories viz. social (9), economic (14) and demographic (5) have been taken and Kenstalls Rank Order Score Method is applied which shows that there is only one taluka in each of very high and very low level of development, while the remaining talukas come in high and medium level of development. The level of agricultural development is measured by taking 16 indicators which have been standardised by the method of per cent proportional standardised mean. The result of the analysis indicates considerable variations in agricultural development. The correlation between regional disparities and agricultural development is worked out with Spearnan's Rank Order Correlation. The *r* values have been calculated for the three group of indicators which have been found to be 94.30 for social, 93.55 for economic, and 88.64 for demographic. It means that high level of regional development leads to high level of agricultural development.

The Bijapur district located in rain-shadow region of western coast has semi-arid climatic conditions. Its economy dwindles due to frequent occurrence of drought and is characterised by underdeveloped economy. Under the Target Area and Target Group Approach of planning this district has been selected and several agricultural development programmes have been inducted into it. **Togarsi** et al. have made an appraisal of these programmes such as Intensive Agricultural District Programme, Intensive Agricultural Area Development Programme. The authors have assessed with reference to the success in HYV, soil and water conservation, manures etc. and have suggested strategies for agricultural development which revolves around the improvement in irrigation system. However, this study gives sufficient information about the weather phenomena, soil conditions, water resources, land holdings, landuse, cropping pattern, agricultural efficiency and socio-economic features of the farm community, livestock and fodder situations. It is hoped that the indepth study undertaken here on various aspects analysed through several appropriate latest techniques will certainly help the agricultural planners to improve agricultural economy in Bijapur district. In order to overcome the drought-proneness, in initiating developmental programmes in such regions, one has to observe how far the strategies of irrigation, animal husbandry and rural industry have succeeded and can suggest developmental strategies. In the above context, agriculture in Bijapur district cannot be improved in isolation, rather it can suitably be developed if other dimensions of the regions such as social, economic and demographic are taken care of to achieve comprehensive and homogeneous development.

The traditional agricultural system in India has undergone considerable changes and conspicuous technological innovations have been noticed after mid-sixties which were initially considered as boon to the national economy but the ultimate result seems to work adversely creating enormous degradation to the total environmental system, both physical as well as social. With this consideration **De** and **Sarkar** have critically reviewed technological changes and their adverse implications, and have suggested alternative strategies for optimal agricultural development keeping in view the ecological implications. They have given an indepth critical analysis of irrigation, improved varieties of seeds, chemical fertilisers and plant protection chemicals, highlighting their affect on environment. While suggesting alternative strategy, they have advocated in depth study of landscape ecology and accordingly development plans compatible to landscape ecology should be introduced which would be, by and large, based on indigenous technology. However, the study concludes that the advanced technology may offer some initial boost to the economy, but ultimately causes enormous environmental degradation, both physical as well as social. Hence, to keep the land and soil permanently productive and to get a steady return out of it for the increasing population, there is a need to have an ecological approach for landscape study first to be followed by the alternative management aspects later. Lastly, instead of getting excessive agricultural products for the time being and loosing everything in the long run, it is better to opt for the alternative strategies which offer steady output regularly keeping the physical as well as social environment intact.

As a result of 40 years planned development programme and policies, not only overall improvement in the economic condition has taken place, but the structure of national and state economies has also changed. But the level of development is not uniform at any level. Inter- and intra-sectoral difference in the economic structure became more sharp and noticeable. So is the case with inter- and intra-regional levels of development. **Mohammad** and **Majeed** have examined the spatial variations in the levels of development in U.P. The study is based on secondary data collected at district level for the year 1986. The level of development of agricultural, industrial social, anthropogenic and economic infrastructure is assessed with the help of 86 indicators which have been aggregated to give overall level of development. All these indicators have been standardised and each sector has been given a weightage score with which the sector averages have been multiplied. The resulting figures of all these averages have been added together and divided by 5 to give composite index of development. Finally the critical analysis of levels of development in all sectors is given separately which indicates a lot of variations into it. The analysis of levels of sectoral and overall development clearly indicates that there is a lot of variations in it which needs to be taken care of while formulating development policies. The districts having lower level of development should be given top priority so that they may come up at par with developed areas and the concept of planning with social justice may be fulfilled.

In an other article **Bhadrapur** and **Naregal** have measured the spatio-temporal level of agricultural development and its correlates in the district of Bellary located in the arid region of India. The levels of agricultural development is measured with the help of 18 indicators by applying Kendall's Ranking Co-efficient Method. Taluka is the unit of analysis and 1975-76 and 1985-86 are the chosen time periods. The study reveals that there is a lot of spatio-temporal variations in the level of development which may be explained in terms of physical, social and techno-economic factors. The results reveal that the environmental factors have positive correlation during 1975-76, whereas out of four factors, only the rainfall is negatively correlated during 1985-86. The demographic factors except per capita net sown area to a cultivator, are positively correlated during 1975-76, whereas during 1985-86 except percentage of agricultural workers to the main workers the remaining factors are positively correlated. During

1975-76 the technological factors except percentage area under commercial crops to net sown area and density of livestock, other factors are positively correlated with the agricultural development, whereas during 1985-86 except density of livestock percentage of electric power consumption and number of pump sets, other factors are positively correlated. The study suggests that all the talukas which are below the group of high development should be considered for immediate process of agricultural development in order of priority. The development of agriculture is examined with the help of rank order correlation technique.

Recent researches on technological change and diffusion of agricultural innovations have revealed that there is only selective adoption of a system of transferred techniques and practices. But very little is known so far about how farmers integrate new and old techniques and practices into intelligible farming system, and even less about how modern scientific ideas interact with traditional system of agricultural knowledge. **Murton** has devoted his research paper to direct attention to the desirability of research-oriented towards an understanding traditional system of agricultural knowledge. Though the cognitive approach with two important connotations : firstly, environmental knowledge and its articulation are internally consistent and self-fulfilling; and secondly, concept may usefully be viewed as local, is applied the focus is on cognition as a critical component in understanding Indian agricultural system. After a brief review of the theoretical background of such an approach, a general outline of Indian system of agricultural knowledge is given. The author has discussed general premises, coordination of agricultural activities, resource predictability and categories of resource discrimination. The study reveals that though changes have begun to manifest themselves with regard to Indian farmer's knowledge and understanding of their environment due to contact with western science by educated elite, but India is a theatre of indescribably complex interaction between the forces of modern science and technology and of an age-old metaphysical tradition. Despite numerous advances, custom is a king not yet and not easily dethroned and his laws are enshrined in the multitude of petty proverbs with which religious songs form a remarkable folk system of knowledge.

No doubt the technological change in agriculture during the last three decades has brought tremendous development in almost all parametres of agriculture which ultimately results in high agricultural productivity but it has posed a number of ecological issues which are now realised with a great concern. **Singh** gives a thorough treatment to these problems in his paper in which he highlights that increase in agricultural production is possible by its horizontal expansion and intensification, and both have adverse effect on environment. It results into land degradation, soil erosion and consequently floods and droughts. The author examines in depth the intensification of agriculture which is mainly possible due to mechanisation, irrigation, HYV of seeds, chemical fertilisers, and pesticides and insecticides. The study after giving a brief account of the positive achievement of all these elements highlights a number of ecological problems which arise out of it. Finally viable suggestions which are based on indigenous technology and suitable for the prevailing environmental condition, have been given. The study concludes that before any innovation is adopted, all possible environmental and human consequences of it must be carefully evaluated and we should be prepared to abandon or not to adopt any technology if it is detrimental to environment or poses any threat to human life.

With the appraisal and experience of the preceding plans, emphasis has kept on changing in the succeeding plans with the aims of bringing overall improvement. For example the result of planning during the first phase of development shows that material benefits accumulated in a few major urban enclaves of economic growth but failed to trickle down to the poor regions. Similarly the adoption of package advanced technology also resulted in uneven growth by favouring better endowed regions and failing to alleviate rural poverty and unemployment in other regions. In the subsequent agricultural development plans emphasis has been given to reduce the regional disparities and elevate the condition of the rural masses. **Karishettar** has examined the agricultural growth and rural poverty. The study shows that with the increase in foodgrain production the percentage of rural population living below poverty line will decline. Similarly, the author has analysed in depth the impact of land reform and irrigation on equitable growth of agriculture and consequently rural upliftment. The study assesses the diffusion of various constituents of Green Revolution — HYV, irrigation, etc. and its impact on employment of human labour which in turn improves the economic condition of the rural people.

India is basically a rural country where about 80 per cent of the population lives in villages. Hence, rurality is the spinal cord of the socio-economic, political and cultural life of India and is the grip of long lasting poverty. Though many efforts have been made to improve the condition of the rurality, the sky remains overcast with few blue patches of development. Various reasons such as limited resources, exploitation of local resources for increasing the national income, sectoral approach in planning, incompetent implementing agencies, urban bias in development, and very much neglected socio-economic structure and change therein (Saini, et al., 1977) are attributed for failure of development programmes. Many of these factors have been emphasised and explained by a number of writers but the factor of the negligence of the social aspects of the rural people which has a great bearing on the fruitful implementation and consequently on the development of the villages still remains unexplored by the planners though it has been given thought off and on. **Singh** and **Pathak** in their paper have examined societal aspects like caste, class and faction and their impact on the development of the villages. They have discussed in detail the nature, form and character of caste, class and faction. In brief the authors have also given a theoretical framework for rural development and have examined and analysed two rural development programmes of India viz. abolition of *zamindari*, and community development programme. Highlighting the village society and problems of development, the authors have suggested two different approaches for the social change: the improvement approach which believes in progressive movement of the society without any radical change in traditional and legal system; and the transformation approach which attempts to establish new forms of agricultural and social organisation and which makes a radical break-through in the existing peasant system in terms of scale of operation, production techniques and social legal structure (Norman Long, 1977).

REFERENCES

Kristensen, T., 1974 : *Development in Rich and Poor Countries*, New York.

Mohammad, Noor, 1976 : Technological Change and Diffusion of Agricultural Innovations, *The Geographer*, Vol. XXIII, pp. 4-14.

———, 1978 : 'Impact of Economic Factors on Diffusion of Agricultural Innovations in Trans-Ghaghara Plain', *Geographical Review of India*, Calcutta.

———, 1978 : 'Impact of Social Factors on Diffusion of Agricultural Innovation in Central Trans-Ghaghara Plain', *National Geographical Journal of India*, Vol. XXIV, pp. 36-46.

Rani, U., 1971 : 'Farm Size and Productivity', *Economic and Political Weekly*, June, pp. A85-A89.

Saini, N. S., et al., 1977 : 'Rural Planning and Development in India' in G.B.K. Rao and N. S. Saini (Eds), *Readings in Urban and Regional Planning in India*, New Delhi.

Sharma, J. S., 1965 : 'Measurement of Agricultural Productivity — Concepts, Definitions etc.' *Journal of the Indian Society of Agricultural Statistics*, No. 1, Vol. XVII, pp. 253-57.

Singh, J., 1972 : 'A New Technique for Measuring Agricultural Efficiency in Haryana', *The Geographer*, Vol. XIX, pp. 14-33.

Sinha, B. N., 1968 : 'Agricultural Efficiency in India', *The Geographer*, Vol. XXI.

Stamp, L. D., 1960 : *Our Developing World*, London, pp. 108-109.

Subbiah, S. and Ahmad, A., 1980 : Determinants of Agricultural Productivity in Tamil Nadu, India," *Transaction of Institute of Indian Geographers*, No. 1, Vol. II, pp. 19-32.

Shafi, M., 1972 : 'Measurement of Agricultural Productivity of the Great Indian Plains', *The Geographers*, Vol. XIX, pp. 4-13.

——, 1974 : 'Perspectives on the Measurement of Agricultural Productivity', *The Geographer*, Vol. XXI, pp. 1-10.

Vendoor, P. J., 1945 : Fattari che Regolano lo Sviluppo Economico della Produttivita del Lavoro, *L'Industrio-I*.

1

THE CONCEPT OF AGRICULTURAL DEVELOPMENT

GOPAL KRISHAN

THOUGH FAIRLY understood in general, the term agricultural development defies a precise definition. Studies on the theme usually skip over the essential of giving a meaning to the term. One looks, in vain, for an elaboration of the concept in the books on agricultural development (Shenoi, 1975) or dynamics of agricultural development (Mohammad, 1979). Implicitly or explicitly, agricultural development is equated with the level and rate of agricultural productivity, in practice crop productivity, in most of the studies.

Is agricultural development synonymous with crop productivity? is the germinal question in this essay. The issue is crucial in the context of growing interest of the scholars in understanding the patterns and processes of agricultural development and of the government in bringing about agricultural transformation on planned lines. After all what is being examined and what is being planned for must be clearly understood. This calls for a comprehensive definition of the term. Some of the hitherto neglected aspects of agricultural development may also get exposed by such an exercise.

The Concept

Agricultural development is unquestionably a multi-dimensional concept of which crop productivity is one of the vital aspects. Crop productivity is to be judged not merely from quantity of production but also from the variety and quality of the produce. The simplest and the crudest measure of crop productivity is the yield per hectare of various crops. A desirable sophistication is introduced by finding out the value of crop produce per hectare of net area sown/cropped area or per cultivator/agricultural worker. Produce per hectare of net area sown or cropped area is an expression of the output per unit of agricultural land, and produce per agricultural worker or cultivator reflects the income level of agricultural population. Sometimes a corrective is applied by computing net output (gross output minus cost of inputs) per hectare of net area sown. This computation involves several methodological problems and is generally given up in favour of a more convenient indicator of output per hectare of agricultural land or per worker.

The variety of agricultural produce can be noted from the corp diversification index. Diversification of cropping as an element of agricultural development is being supported not on economic grounds but on considerations of self-reliance in agricultural production and maintenance of soil fertility. Specialisation in cropping, including mono-culture, may yield rich dividends for a while but mono-culture is inimical to natural order, deleterious to ecospheric safety and lethal to man's long-term interests (Igbozurike, 1971).

The quality of crop produce has seldom been taken into account in studies relating to agricultural development. The quality of crops does differ and carries its implications for the nutrient value, feeding capacity and taste in the case of food crops and for the shine, efficiency, durability and convenience of processing in the case of industrial crops. A practical way of ascertaining the quality of the produce would be to take a note of the variety of seeds being used for various crops like wheat, rice, cotton, sugarcane and vegetables.

Diversification of agriculture must be construed as the second vital aspect of agricultural development. It stands for promotion of dairying, cattle-rearing, poultry-farming, pig-raising, bee-keeping and fishing. These activities not only significantly contribute to the basket of agricultural products but also generate additional employment for the surplus agricultural populations, especially in crowded agricultural countries like India. Suitable indicators for measurement of diversification of agriculture can be adopted. These may include the proportion of workers in these activities to total workers in primary activities and the contribution of these activities in the total agricultural produce. In the absence of requisite data, diversification of agriculture may be inferred from the number of livestock units per 100 hectares of net area sown, the number of poultry birds per 100 persons and the ratio between milch cattle and population. Efforts at diversification of agriculture may be concentrated more in the case of small farmers for whom these activities can open new avenues of gainful employment.

Commercialisation of agriculture is still another dimension of agricultural development. The degree to which market forces have penetrated an area and the scale upon which they operate will be the crucial factors in almost every question of agricultural development (Hunter, 1969). A directed effort at commercialisation of agriculture may involve articulation of farmers' psychology toward that end, provision of marketing facilities and above all a rise in the yield of agricultural produce. The percentage of cropped area under cash crops may be used as a measure of commercialisation of agriculture. The density of market centres per 1000 sq. km. of area can also provide a clue to the degree of commercial agriculture. A more reliable indicator would be the share of market arrivals in total production both of which can be computed in terms of their cash value.

The development of agriculture is to be judged also from the degree of equity in farm incomes and the nature of agrarian relations. The test of any development is the extent to which it delivers social justice. The unequal distribution of land ownership and economic power makes it possible and profitable for the landowner to combine various modes of exploitation of the rural poor (Davey, 1975). One of the conspicuous weaknesses of the Green Revolution was that it widened the disparities in farm incomes (Frankel, 1971). A rethinking on development strategy became necessary. It was realised that to make the gains of the new revolution equitable, greater stress should be put on biochemical inputs (fertilisers, irrigation, and high-yielding variety of seeds) and less on mechanical inputs (tractors and harvesters). The methods of cultivation and the technology used mirror the nature and level of agricultural development attained by an area.

A study of disparities in farm incomes would require collection of primary data. The concentration index of landholdings can be used as a surrogate. The distributional pattern of assets can be adopted as another indicator to assess disparities within an agricultural population.

Non-exploitative and tension-free agrarian relations are an essential ingredient of any agricultural development. Relations between big farmers and small farmers as well as between farmers and landless labourers are under reference. The nature of relations between different categories of farmers may be inferred from the comparative benefits they are deriving from the various decision-making bodies in respect of supply of irrigation water, fertilisers, high-yielding variety of seeds, loans, marketing facilities and land reform policies. The farmer-labourer relations get reflected in the wage rate which the labourer is getting. This indicator does give an idea of the justice that the labourers are getting at the hands of cultivators but may not suggest the degree of harmony prevailing between them. Tensions between the two can be high even under conditions of high agricultural wage rates. Indeed empirical studies of agrarian relations in different areas would be rewarding and will help in identifying dependable indicators on the matter.

Above all agricultural development should not produce deterioration in ecological conditions. It should not lead to defacement of forests, exhaustion of soil nutrients, depletion of underground water and emergence of waterlogging conditions. Conservation of physical resources is an integral part of any agricultural development.

It follows that agricultural development is a much more comprehensive concept than normally understood. Productivity is only one of its dimensions. Agricultural development, in a true sense, denotes the quality of the agricultural system of a region in terms of productivity, diversification and commercialisation consistent with a desired state of agrarian relations and ecological balance. The level and the rate of agricultural development may also be distinguished: the former represents a picture prevailing at a particular point in time while the latter stands for the progress achieved over a given period. If the process of agricultural development is regulated on systematic lines, it becomes agricultural development planning.

Agricultural Development in Geographical Studies in India

A review of geographical literature covering agricultural development in India reveals that seldom an attempt is made to define agricultural development and to select criteria in the light of any conceptual framework. Criteria are chosen on an ad hoc basis depending upon convenience of data availability. No attempt is made to explain the significance of criteria used. Rarely a distinction is made between the elements of agricultural development and the factors in agricultural development. Focus on productivity dimension is dominant. Only a few studies take into account the matters relating to commercialisation of agriculture, disparities among agricultural populations and nature of agrarian relations.

In the mapping of the levels of economic development in India, Schwartzberg included in his exercise the agricultural indicators of crop productivity (average annual food production per capita of agricultural population and per-acre yield of major foodgrains as percentage of national yield), agrarian relations (agricultural labourers as per cent of rural population, average annual wage per agricultural labour family and concentration ratio of land holdings) and institutional factor of members of agricultural credit socictics per million of agricultural families (Schwartzberg, 1962). These indicators were used not to measure specifically the levels of agricultural development but were listed among the structural, productivity and consumption-cum-welfare indicators of economic development. Nevertheless an idea is obtained as to what the author considered as vital from the viewpoint of agricultural development.

Mitra was more specific in the selection of agricultural indicators of regional development (Mitra, 1967). IIis indicalors included percentage of double cropped area as an expression of intensity of cultivation, percentage of gross irrigated area as a measure of agricultural inputs, area under cash crops signifying commercialisation of agriculture and yield of clean rice indicating efficiency of agricultural practices. Other indicators were the percentage of households cultivating 0-5 acres as an index of the degree of subsistence economy, percentage of pure tenancy holdings to obtain an idea about land dependency of cultivators, percentage of hired workers as a measure of market orientation of agriculture, percentage of agricultural workers indicating dependence of a population on this activity and number of agricultural workers per 100 acres of net area sown to assess pressure on agricultural land. Evidently Mitra selected some indicators (such as percentage of households cultivating 0-5 acres and percentage of pure tenancy holdings) which were at cross with agricultural development. With appropriate adjustment of data (e.g. taking percentage of households cultivating 5+ acres in place of percentage of households cultivating 0-5 acres), he could have made all the indicators positive. Anyhow Mitra's focus remained primarily on physical and social infrastructure while assessing the agricultural perfonnance of different areas.

A composite index of agricultural development based on three factors — growth rate of agricultural output, use of modern inputs in agriculture and productivity per hectare — was constructed by Nath (1962) to compare the level of agricultural development in various States of India. His was one of the few papers which was directly concerned with the spatial patterns of agricultural development in

India. However, he confined his attention mainly to the productivity component of the development and examined spatial variations only up to the level of individual States.

Sharma used detailed districtwise data for a study on the same theme (Sharma, 1971). He opined that agricultural development should be assessed not only by levels of productivity or trends in agricultural production but also with reference to various physical inputs like irrigation, fertilisers, improved seeds and extent of cultivated area. He regionalised India on the basis of land resources (including average annual rainfall, per capita gross area sown, percentages of gross irrigated, net sown and double-cropped area, and soil fertility) and crop productivity index. Although the author did not cover all the dimensions of agricultural development in his regionalisation he rightly stressed the need of devoting due attention to the factor of land resource potential in such attempts.

Using data for individual tehsils in his study of regional disparities in development in Andhra Pradesh, Alam employed six indicators for the agricultural sector (Alam, 1974). Two of these related to productivity (agricultural output per agricultural worker and per acre) and four to factors in agricultural development (percentages of gross irrigated, gross cropped, canal irrigated and double-cropped area). Indicators pertaining to commercialisation and diversification of agriculture as also to agrarian relations were distinctly missing.

In his remarkably analytical study of levels of regional development in India, Raza took as many as 41 indicators of agricultural development grouped into four subsets of productivity, production conditions, agrarian relations and change in agriculture (Raza, 1978). Productivity was measured in terms of output per hectare, per worker, per tractor, per tube-well and per unit of fertiliser. Production conditions included percentages of cultivated area and irrigated area, male workers per 100 hectares, tractors, tube-wells and fertiliser units per 10,000 hectares, percentage of area under commercial crops, cropping intensity, implements per 10,000 hectares and per household, and farm business investment per cultivator household. Agrarian relations were judged from the ratio between agricultural labourers and cultivators, percentages of households owning .01 to 5 acres and between 5 and 15 acres in rural areas, and Ginni coefficients of productive assets, total assets and operational landholdings. Change in agriculture was measured in terms of increase during 1961-71 in male agricultural labourers per 100 hectares of net area sown, value of output, intensity of cultivation, proportion of irrigated area, number of tube-wells and units of fertilisers used. While one admires the extensive range of indicators put into service one fails to understand the logic of including indicators like output per tractor, per tube-well or per unit of fertiliser. Why the indicators of percentage of households owning land between .01 and 5 acres and between 5 and 15 acres in rural areas were taken simultaneously while they function opposite to each other? How does the number of male workers per 100 hectares (which is an index to the pressure of population on land resources) reflect agricultural development? Such discrepancies arose mainly because the selected criteria were neither critically examined nor coherently arranged. A brief justification of the indicators should have enhanced the quality of the work which is otherwise extremely stimulating in description and interpretation of the spatial patterns of agricultural development in India.

The above review of the selective, but sufficiently representative, literature testifies that all the studies take a partial view, in varying degrees, of agricultural development. The main attention was devoted to the dimension of productivity and its determinants. All the studies were constrained by the problem of data availability. The hunger for data was so voracious that even statistics of marginal value were not spared from usage and that too without any critical evaluation.

Agricultural Development and the Geographer

Agricultural development can be a fascinating field of study in geography. The geographer's role lies primarily in identification, description and interpretation of regional variations in the level and the rate of agricultural development. And this must be done in the context of physical resource base, technological advancement, institutional framework and external relations of different areas. Secondly, the spatial diffusion of agricultural development in its nature, intensity and direction presents another interesting field of investigation. The basic geographic query would be as to how the different areas respond to the impulses of development in agriculture. Lastly, geographers can provide a typology of different areas from the viewpoint of agricultural planning. Such a regionalisation must have a multi-dimensional perspective imbibing existing patterns, trends and potentialities of agricultural development.

REFERENCES

Alam S., Manzoor (ed.), 1974 : *Planning Atlas of Andhra Pradesh*, Pilot Map Production Plant, Survey of India, Hyderabad, p. 3.

Davey, Brian, 1975 : *The Economic Development of India*, Spokesman Books, p. 175.

Frankel, Francine R., 1971 : *India's Green Revolution*, Princeton University Press, Bombay, p. 191.

Hunter Guy, 1969 : *Modernising Peasant Societies*, Oxford University Press, London.

Igbozurike, Matthias U., 1971 : "Against Mono-culture", *Professional Geographer*, 23, p. 114.

Mitra, Asok, 1967 : *Levels of Regional Development in India*, Indian Statistical Institute, New Delhi, pp. 8-9.

Mohammad, Ali (ed.), 1979 : *Dynamics of Agricultural Development in India*, Concept Publishing Company, New Delhi.

Nath, V., 1969 : "The Growth of Indian Agriculture : A Regional Analysis", *Geographical Review*, 59. p. 369.

Raza, Moonis, 1978 : *Levels of Regional Development in India*, Indo-Soviet Symposium on Regional Development and National Planning, Tiblisi-Baku, Appendix II.

Schwartzberg, Joseph E., 1962 : "Three Approaches to the Mapping of Economic Development in India", *Annals of the Association of American Geographers*, 52, p. 462.

Sharma, P. S., 1971 : "Agricultural Regionalisation of India", in A. Chandra Sekhar (ed.), *Economic and Socio-Cultural Dimensions of Regionalisation*, Registrar General, New Delhi, pp. 253-278.

Shenoi, P. V., 1975 : *Agricultural Development in India*, Vikas Publishing House, Delhi.

2

CORRELATES AND DETERMINANTS OF AGRICULTURAL PRODUCTIVITY GROWTH

A Test of Some Hypotheses

EMILIO CASETTI

THE GROWTH of labour productivity has a crucial bearing upon two major issues of our time: inflation and development. The increase in labour productivity is not just a correlate of economic development but possibly its single most important dimension. Also it constitutes one of the major factors with the potential to counteract the inflationary pressures that make their presence felt in just about any society throughout the world. In fact, increases in the remuneration of the factors of production are not inflationary only to the extent that in the aggregate they balance productivity increases.

This paper is concerned with testing some hypotheses about the dynamics of labour productivity in agriculture. Specifically, cross-sectional State-level data on labour productivity changes in agriculture are used to seek answers to the following questions:

(a) Is the increase in the labour productivity in agriculture positively related to the rate of growth of agricultural output in the manner suggested by the "Verdoorn law" ?

(b) Is there a tendency for the agricultural labour productivity to grow faster in countries in an intermediate position on the development continuum? In this paper this tendency will be referred to as the "Kristensen effect".

(c) Does the labour productivity in agriculture progress more quickly in countries that experience greater population pressures?

In the sections that follow, first the pertinent theoretical issues will be brought into focus. Then, the data set, the hypotheses tested, and the results obtained will be presented and discussed.

Verdoorn Law

Verdoorn (1949) suggested that there is a positive linear relation between the rate of growth of productivity and the rate of growth of output. His statement was based on data for manufacturing as a whole and for individual industrial sectors in different countries. Also most of the subsequent attempts to verify Verdoorn law have been based on manufacturing data (Salter, 1960; Kennedy, 1971; Cripps and Tarling, 1973). However, a variant of Verdoorn law sometimes referred to as Kaldor law has been tested for other economic sectors as well. Kaldor law relates growth rates of labour productivity to growth rates of employment rather than of output (Kaldor, 1966, 1967; Cripps and Tarling, 1973; Rowthorn, 1975; Cornwall, 1976).

The empirical tendency for expanding economic sectors to experience higher rates of productivity growth can be explained by some combinations of the following: (1) expanding sectors may benefit from external economies and from economies of scale, (2) economic growth is associated with an inflow of capital with embodied technological progress, (3) the most productive workers who are carriers of modernity values and are trained in the latest technologies find greater opportunities for employment in expanding economic sectors than in sectors that stagnate or decline, (4) modern technologies and the "learning by doing" that goes with them is more likely to start within the expanding segments of an economic sector but eventually diffuses throughout the entire sector.

These factors of productivity growth may differ in intensity and manner of operation in various circumstances, but there seems to be no ground for presuming that they are restricted to manufacturing. An investigation of whether Verdoorn law is applicable to the agricultural sector is one of the points of focus of this paper.

The Kristensen Effect

A tendency of the countries at the intermediate level of development to experience higher rates of economic growth than those of the countries closer to the end points of the development continuum, has been noted by some authors (Russett et al, 1964, p. 309 ff.; Kristensen, 1974, p. 24 ff.; Casetti, 1979) and questioned by others (Hagen and Hawrylyshyn, 1969, pp. 88-89), However, the weight of the available evidence seems to support it.

This tendency has been linked to the differences in technological absorption capabilities at different levels of economic development. The least developed societies, due to their socio-economic structure, and because of a lack of capital and trained manpower, have a limited ability to apply the existing stock of scientific and technological knowledge. As they develop, their ability to use existing technologies, grows, while at the same time the stock of applicable technologies unused by them dwindles, thus slowing down their rate of growth. In the limit, the economic growth becomes dependent upon the creation of new scientific and technological knowledge as it is the case for the developed countries.

Recently these propositions were formalised and tested with respect to their efficacy for explaining empirical relations between contemporary rates and levels of economic development (Casetti, 1979).

Some aspects of this conceptual framework have been applied to the study of productivity (Gomulka, 1971). The hypothesis that the growth of labour productivity in manufacturing for developed countries is very much dependent on the technological gap between an industrial leader and the country in question has been tested in a recent U.N. study* and by Cornwall (1976).

In this paper the hypothesis that the highest growth rates of agricultural labour productivity tend to occur in countries at intermediate levels of development will be tested.

Population Pressure and Productivity Growth

In recent years the question whether population "pressures" have a favourable impact on economic development has been the object of controversies based on theoretical propositions with a potential relevance to labour productivity analyses (Robinson and Horlacher, 1971; Simon 1977; Clark, 1978). Population pressure is sometimes operationalised in terms of high population density, taken to imply a high ratio of population to resources; and sometimes as high rates of

---

* *U.N. 1970,* Ch.3.

population growth. The implications of each of these have not been clearly separated, and at any rate the population optimists have tended to view both in positive terms. It has been argued that population pressures stimulate innovation and technical progress and, in general, are conducive to get individuals and nations to work harder and to innovate more. Also, it has been claimed that denser and faster-growing populations benefit from scale economies. On the opposite side of these issues, negative effects of high population pressures have been associated with diseconomies of scale and external diseconomies due to congestion, pollution and resources depletion. Also it has been claimed that high levels of population growth are associated with lower savings and at the same time tend to channel a greater portion of available savings to "demographic" investment as opposed to the investments with a stronger impact on economic progress (Coale and Hoover, 1958). The empirical analyses aimed at identifying whether a faster population growth is associated with higher rates of development remained inconclusive (Easterlin, 1967; Chesnais and Sauvy, 1973; Simon, 1977) Instead, recent investigations into the effects of population density on development seems to be consistent with the optimists' views (Szymanski, 1974; Stryker, 1977).

The theoretical propositions regarding the effects of population pressure can be translated into statements concerning the determinants of labour productivity in agriculture. In the agricultural sector, however, there is a complicating factor that needs to be taken into account. This factor is the movement of population out of agriculture that has been long since recognised as a concomitant of economic development. Since it can be presumed that in the balance the people who move out of agriculture are the ones that provide the least productive labour inputs to it, and since the exodus of people from the countryside is associated with increases in farm size that facilitate technological progress, it ought to be presumed that in agriculture growth of rates of labour force are associated with lower rates of growth productivity (Kumar, 1973, p. 182 ff.; Hayami, 1971). But the rate of growth of the agricultural labour force is related to the rate of growth of population, which in turn implies that the gross effect of the rates of growth of population on productivity can be negative, even if the corresponding net effect is positive. This net effect is the effect of population growth on the productivity growth when the rate of change of agriculture labour force is controlled.

In this paper the hypothesis that population density and rates of growth of population are positively related to growth rates of agricultural productivity is tested.

The Data

The data used in this investigation pertain to the following variables:

(1) Percentage rate of change of population to be denoted by P^*. All the per cent rates of change in the data set are per year and refer to the decade 1960-70.
(2) Percentage rate of change of labour force in agriculture, denoted by LA.*
(3) Percentage rate of change of GDP originating in agriculture, denoted by YA^*.
(4) Population density in people per square kilometre in 1970, denoted by D.
(5) GNP per capita in the year, 1965, in U.S. dollars, denoted by y.
(6) Percentage rate of change of labour productivity in agriculture, denoted by A^*.

P^* and YA^* were obtained by averaging the rates for the year 1960-65 and 1965-70, published in *World Bank* (1976, p. 392 ff.),

LA^* was derived from data on the percentage of labour force in agriculture and total labour force for the years 1960 and 1970 (*World Bank 1976,* p. 514 ff.) Specifically, labour force data in agriculture for 1960 and 1970 were first obtained [LA (1960)] and [LA (1970)] and then LA* was generated from them using the following equation: LA*=(.1)1n[LA(1970)/LA(1960)]. y is from Taylor and Hudson (1972, p. 314 ff,). Finally, A* was calculated from YA* and LA* using the definition $A^* = YA^* - LA^*$.

The initial data set included all countries with a 1973 population of 2 million people or more, then all observations with missing values on one or more variables were removed. Also removed were Saudi Arabia and Singapore respectively because the oil revenues in the first case and the exceptionally high population densities in the second render these countries not representative of a "normal" situation. The data employed in the analysis includes 71 observations and seven variables.

* Indicates percentage rate of interest throughout.

Hypothesis Testing

The regression equations used to test the occurrence of Verdoorn, Kristensen and population pressure effects are shown in Tables 2.1 and 2.2. Table 2.1 refers to the three effects considered in isolation. Instead the regressions in Table 2.2 test the concurrent operation of two effects or more.

In all the regression equations the percentage rate of change of agricultural productivity constitutes the dependent variable. The independent variables are various combinations of the variables associated with the three effects. The "effects" are tested by determining whether the regression coefficients of the variables to which each effect is associated have the requisite sign and are significant at the five per cent level or greater on the basis of their *t* values. In the tables, the *t* values are in parenthesis under the respective correlation coefficients.

The hypothesis that Verdoorn law operates in the agricultural sector was confirmed. The regression number 1 in Table 2.1 shows that the growth rates of agricultural output are significantly related to the growth rates of agricultural labour productivity. However, the variance explained by this regression is only six per cent, which indicates that the Verdoorn effect shows up, but weakly.

The hypothesis that the growth of agricultural labour productivity A^* tends to be highest at intermediate levels of development is tested in Equation 2 of Table 2.1. In this equation A^* is regressed against a quadratic function of the GNP per capita variable y. Both the coefficient of the linear and of the quadratic terms of the equation are significant. It can be easily shown that the estimated equation predicts a maximum percentage growth rate of agricultural labour productivity of about 5.8 per cent at a GNP per capita of 1709 dollars. The values of GNP per capita for the countries in the data set range from a minimum of about 50 dollars to a maximum of about 3,600 dollars. The USA values of A^* predicted by the regression equation for these y's are respectively two and nine per cent. Consequently, the hypothesis that the highest levels of rates of growth of productivity tend to occur at intermediate levels of GNP per capita is consistent with the data.

The population pressure effects are tested in Equation 3 through row 7 of Table 2.1. The effect of population density D on rates of growth of agricultural productivity proves significant but it has a very weak explanatory power, since it accounts for only eight per cent of

TABLE 2.1: **Regression tests for Verdoorn, Kristensen and population pressure effects**

Number	Constant	y	y^2	D	P^*	LA^*	YA^*	R	R^2
(1)	2.1330	—	—	—	—	—	.4188 (2.19)	.25	.06
(2)	1.7645	4.7069 (4.53)	–1.3765 (–3.58)	—	—	—	—	.52	.27
(3)	2.8738	—	—	8.6182E-3 (2.45)	—	—	—	.28	.08
(4)	5.5459	—	—	—	–.9544 (-3.23)	—	—	.36	.13
(5)	3.3276	—	—	—	—	–.7968 (-11.92)	—	.82	.67
(6)	1.0543	—	—	—	1.0363 (4.73)	–1.0618 (-13.11)	—	.87	.75
(7)	.9381	—	—	1.1977E-3 (.60)	1.0504 (4.74)	–1.0525 (12.72)	—	.87	.75

Legend:

y = GNP per capita.

D = population density, in population per square km.

P = yearly percentage of population.

LA^* = yearly percentage rate of change of the agricultural labour force.

YA^* = yearly percentage rate of change of agricultural product.

t values are in parenthesis under the respective regression coefficients.

the variance of A^*. Equation 4 shows that the gross effect of the rate of growth of population P^* on the rate of growth of productivity is significant and negative. This result would seem to support the pessimist school of thought regarding the effects of population growth. However, when the rate of growth of the agricultural labour force LA^* is entered into the regression (Equation 6) the regression coefficient of the rate of growth of population is significant and positive in sign. This can be taken to mean that the high rates of population growth exercise a positive influence on the productivity growth when the effect of the rate of change in agricultural labour force is controlled. The sign of the regression coefficients of LA^* is negative in Equation 6 and also in Equation 5, in which LA^* is the only independent variable. Interestingly enough, LA^* is the single greatest predictor of A^* among the variables used in this analysis, and by itself explains 67 per cent of the variance of agricultural productivity growth. Clearly the removal of the least productive agricultural workers, and the increase in farm size presumably associated with declines of agricultural labour force (or with lower growth of it) constitute the greatest correlates of labour productivity growth in agriculture.

When population density, rate of population growth, and rate of agricultural labour force are simultaneously entered into the regression equation (Equation 7, Table 2.1) the effect of population density remains positive but is no longer significant, while the regression coefficients and values of P^* and LA^* are almost identical to their counterparts in Equation 6.

In conclusion, these tests suggest that both population density and population growth exercise a positive influence on the rate of growth of agricultural productivity. However, both effects are comparatively weak, and the positive net effect of high rates of population growth show up only when the dominant effect of the rate of change of the agricultural labour force is controlled.

In summation, the analyses described so far show that the rate of growth of agricultural labour productivity tends to be positively related to the growth rate of the agricultural sector, tends to be greater at intermediate levels of development, and is positively affected by population pressure when the rate of change of labour force is controlled. These effects, however, are not independent from one another, since they all reflect, albeit in different manner and to a different degree, mechanisms such as technological progress, scale economies, external economies and learning by doing. This raises the question of whether each effect is still significant when the variables associated with other effects are controlled. For example, is the growth rate of the agricultural sector still positively related to the growth rate of labour productivity, when the effect of population pressures on agricultural productivity is controlled? Or, do still countries at an intermediate level of development tend to experience higher rates of productivity growth when the Verdoorn effect is controlled? And so on.

The regression equations testing for the simultaneous occurrence of more than one of the three effects are given in Table 2.2. Specifically, Equations 1, 2, and 3 include variables associated with population pressure effects and with the tendency for highest progress to occur at intermediate development levels. Equation 4 tests joint Verdoorn and Kristensen effects. Equation 5 tests joint population pressure and Verdoorn effects, and finally Equation 6 incorporates simultaneously variables associated with all the three effects.

In Table 2.2 the variables associated with the same effect are placed in contiguous columns. Variables y and y^2 always appear in pairs, and so do variables P^* LA^*. The population pressure effect is represented either by the variable D or by the variable pair P^* and LA^*. LA^* and YA^* cannot appear in the same regression equation since A^* is the dependent variable and by definition $A^* = YA^* - LA^*$. Consequently, only the population density variable D is used to represent population pressure effects whenever YA^* appears in a regression equation.

The three effects manifest themselves in all the regressions in Table 2.2, in the sense that in every single instance the signs of the regression coefficients are consistent with the corresponding ones in the analyses in which the Verdoorn, Kristensen, and population pressure effects were considered in isolation. On the other hand, adding independent variables to the regression equations to test simultaneously which the occurrence of more than one effect causes some regression coefficients to appear, are not significant. Yet, in the balance, the equations in Table 2.2 suggest that each of the three effects influences to some extent the rates of growth of agricultural labour productivity when the other effects are controlled.

* Percentage rate.

TABLE 2.2: Regression tests for the joint occurrence of Verdoorn, Kristensen and population pressure effects

Number	Constant	y	y^2	D	P^*	LA^*	YA^*	R	R^2
1.	1.6120	4.2961	–1.2459	3.986iE-3	—	—	—	.53	.29
		(3.94)	(–3.13)	(1.20)					
2.	1.1819	.4248	–.2980	6.0449E-4	.9709	–1.0735	—	.88	.77
		(.55)	(1.19)	(.31)	(4.24)	(–11.69)			
3.	1.2334	.4543	–.3100	—	.9639	–1.0773	—	.88	.77
		(.60)	(–1.26)		(4.26)	(–11.93)			
4.	–1.4562	5.3355	–1.3359	—	—	—	.8347	.69	.48
		(5.97)	(–4.08)				(5.16)		
5.	1.2541	—	—	9.5419E-3	—	—	.4729	.40	.16
				(2.80)			(2.58)		
6.	–1.6665	4.8753	–1.1868	4.5342E-3	—	—	.8443	.71	.50
		(5.25)	(–3.53)	(1.61)			(5.28)		

Note: See legend of Table 2.1 for explanation of symbols.

Conclusion

Cross-sectional data for a sample of 71 countries were used to test the following hypotheses: (1) that the growth rates of agricultural labour productivity are positively related to growth of agricultural output, (2) that productivity grows faster in countrics at intermediate levels of dcvelopment, and (3) that population pressures tend to be associated with higher rates of growth of labour productivity. All of these hypotheses were supported by the data.

REFERENCES

Casetti, E., 1979 : "A Class of Differential Equations Relating Rates and Levels of Economic Development", *Proceedings of the Tenth Annual Pittsburgh Conference on Modelling and Simulation,* 10, pp. 1419-1423.

Chesnais, J. C. and Sauvy A., 1973 : "Progress Economiques et Accroissement de la Population : une Experience Commentee", *Population* 28, pp. 843-57.

Clark, C., 1978 : "Population Growth and Productivity", Research in Population Economics 1, pp. 143-154.

Coale, A. J. and Hoover E. M., 1958 : *Population Growth and Economic Development in Low Income Countries,* Princeton University Press, Princeton.

Cornwall, J., 1976 : "Diffusion, Convergence and Kaldor's Laws", *Economic Journal,* 86, pp. 307-14.

Cripps, T. F. and Tarling R. J., 1973 : *Growth in Advanced Capitalist Economies 1950-1970,* Cambridge University Press, Cambridge.

Easterlin, R. A., 1976 : "Effects of Population Growth on the Economic Development of Developing Countries", *The Annals of the American Academy of Political and Social Science,* 396, pp. 98-108.

Gomulka, S., 1971 : *Inventive Activity, Diffusion, and the Stages of Economic Growth,* Aarhus.

Hagen, E. E. and Hawrylyshyn, O., 1969 : "Analysis of World Income and Growth", *Economic Development and Cultural Change,* 18, pp. 1-96.

Hayami, Y. and Ruttan V. W. 1971 : *Agricultural at Development : An International Perspective,* The Johns Hopkins Press, Baltimore, Maryland.

Kaldor, N., 1966 : *Causes of the Slow Rate of Growth in the United Kingdom,* Cambridge University Press, Cambridge.

———, 1967 : *Strategic Factors in Economic Development,* New York State School of Industrial and Labour Relations, Cornell University, Ithaca, New York.

Kennedy, K. A., 1971 : *Productivity and Industrial Growth : The Irish Experience,* Clarendon Press, Oxford.

Kristensen, T., 1974 : *Development in Rich and Poor Countries,* Praeger Publishers, New York.

Kumar, J., 1973 : *Population and Land in World Agriculture : Recent Trends and Relationships, Population Monograph Series No. 12,* Institute of International Studies, University of California, Berkeley, California.

Robinson, W. C. and Horlacher, D. E., 1971 : *Population Growth and Economic Welfare, Reports on Population Family Planning,* 6, pp. 1-40.

Rowthorn, R. E., 1975 : "What Remains of Kaldor's Law?", *The Economic Journal,* 85, pp. 10-19.

Russett, B. M., and Alker H. R. Jr., Deutsch K. W. and Lasswell H. D., 1964 : *World Handbook of Political and Social Indicators,* Yale University Press, New Haven, Connecticut.

Salter, W. E. G., 1960 : *Productivity and Technical Change,* Cambridge University Press, Cambridge.

Simon, J. L., 1977 : *The Economics of Population Growth,* Princeton University Press, Princeton, N. J.

Stryker, J. D., 1977 : "Optimum Population in Rural Areas : Empirical Evidence from the France Zone", *The Quarterly Journal of Economics,* 91, pp. 177-93.

Szymanski, A., 1974 : "Economic Development and Population", *Studies in Comparative International Development,* 9, pp. 53-69.

Taylor, C. L. and Hudson, M. C., 1972 : *World Handbook of Political and Social Indicators,* Yale University Press, New Haven, Connecticut.

United Nation, 1970 : *Economic Survey of Europe 1969,* Part I, New York.

Verdoorn, P. J. "Fattori che Regolano lo Sviluppo Economico della Producttivita del Lavoro", *l' Industria,* 1.

World Bank, World Tables 1976, The Johns Hopkins University Press, Baltimore, Maryland, 1976.

3

NATURAL RESOURCES IN ECONOMIC GROWTH*

S. V. CIRIACY-WANTRUP

DURING THE past 10 years, it has become the fashion among economists to emphasise the decreasing importance of natural resources in economic growth. This applies especially to the renewable resources used by agriculture (Schultz, 1960, 1968). Use of these resources is strongly influenced by economic institutions, both in developed and undeveloped economies. It is not surprising, therefore, that there has been, in recent literature, a de-emphasis of the role of institutions paralleling that of natural resources. In the current mathematical models of economic growth, the role of institutions is not explicitly considered (Durkheim, 1893; Jorgenson, 1961; Fei, 1964). In a well-known book on transformation traditional agriculture (Schultz, 1964), the author devotes only two pages to the influence of economic institutions on land tenancy.[1]

Emphasis has shifted from natural resources and economic institutions to technological change or, in production function terminology, to "new" or "modern" factors of production. If institutions are considered at all, they are treated as factors furnishing services like other factors of production. Changes in the kind and quantity of institutional services are regarded as determined by an economic demand-supply scheme (Schultz, 1968).

In accordance with my assignment, I should like to focus on this doctrine of the decreasing importance of natural resources and the relatively insignificant and passive role of institutions in economic

* Giannini Foundation Research Paper 294. Reprinted from the *American Journal of Agricultural Economics*, Vol. 51, 1969, as desired by the Author.

growth. I should like to analyse its validity, inquire into its relevancy, and show its implications or policy in developing countries.

Decreasing Importance of Natural Resources in Economic Growth ?

The doctrine of the decreasing importance of natural resources in economic growth is based on the observation, documented largely for the United States, that the shares of agriculture and of agricultural land in the national income have decreased in the course of economic growth. Two questions arise with respect to this observation.

First, does the fact that the income shares of agriculture and agricultural land have decreased in the United States in the last quarter of the 19th century mean that the income shares of all natural resources used by agriculture, including those used also by the rest of the economy, have decreased ? Further, can the experience with the agricultural sector in the United States during a particular period of economic history be extrapolated to other sectors, other time periods, and other national economies?

Second, even if the first question can be answered in the affirmative, what is the meaning of a decreasing income share in terms of the "importance" of natural resources in economic growth? What are the implications for explaining, projecting, and influencing economic growth?

Let us turn to the first question. The natural resources used by agriculture are mainly land, water and climate. All three are used also by industry, transportation, urban settlements and recreation. For this reason, referring to differences in demand elasticities for the products of agriculture on the one side and those of the industrial, transportation, urban and recreational sectors on the other and then equating agriculture with natural resources are not permissible.

Shifts of these natural resources out of agriculture into other uses proceed at different rates and with different geographical impacts. In California, for example, the shifts of agricultural land and water into industrial transportation, urban and recreational uses is a relatively recent phenomenon but has progressed further than in other parts of the country (Ciriacy-Wantrup, 1964). This shift involves price increases for natural resources that are frequently a multiple of prices prevailing in agriculture. Admittedly, the quantities of land and water used in agriculture are still large compared with those used in other sectors.[2]

But this relationship is changing in favour of non-agricultural uses.

Next, we may note that the decrease in the income share of agriculture in the United States took place in a period of history when the terms of trade moved against agriculture internationally for a number of specific historical reasons which are not likely to recur in the same combination. Terms of trade were affected by the expansion of agriculture into less-developed regions, the mechanisation and intensification of agriculture in more developed regions, and, cyclically, by economic fluctuations—especially during the 50-year "cycle" connected with the name of Kondratieff (Ciriacy-Wantrup, 1936). Cyclical effects are distorting when income shares of sectors and factors are compared for periods, such as census years, falling into different phases of economic fluctuations.

With respect to climate and the increasingly important group of amenity resources—such as scenic attractions, public parks, clean water and air, and wilderness areas—it is difficult or impossible to evaluate shares in national income and wealth because the market system furnishes price only incompletely or not at all. Still there is little doubt that the significance of these resources for the location of industry and non-agricultural residence—including second homes—and for the income shares going to transportation, communication, and other services has greatly increased in the United States during the same period for which a decrease in the importance of natural resources is being claimed. There is also little doubt that the price society must pay for safeguarding the quality of these resources is increasing greatly and will continue to do so in the future.

With respect to natural resources originating and mainly used outside of agriculture, such as minerals and energy resources, there is evidence that their income shares have increased in the United States in some periods and decreased in others (Potter and Francis, 1961). Here also the effect of economic fluctuations is significant.

In summary, then, we should be cautious in extrapolating an observation that applies to agriculture and agricultural land in the United States during a particular period of its economic history to all natural resources, time periods, and economics. On the basis of the available evidence, the validity of such extrapolation appears at least questionable. The possibilities for such extrapolation depend on the type of natural resource considered, on the type of technological change that interacts with the resource, on factors affecting the terms of trade of the products of the resource, and, last but not least, on the degree to which the resource is integrated into the market system.

Let us now turn to our second question and disregard for a moment the cautionary remarks just made. What does the alleged decreasing "importance" of natural resources in economic growth mean for explaining and projecting economic growth and for public policy aimed at generating and influencing economic growth? The reply, I am afraid, must be that it is irrelevant.

A decreasing income share may be due to greater efficiencies in a sector favoured by technological change, making its product cheaper—especially if demand elasticities are low—and/or releasing factors for other employment. Are we to argue that such a sector or the services employed in it have become less important for economic growth? The opposite would seem more plausible.

A decreasing income share of a sector, on the other hand, may be due to the growth of other sectors producing substitutes for particular uses. Historically, this has been the case in many minerals and energy resources (charcoal, bituminous coal, sperm oil, etc.). Still, this does not indicate that the individual natural resources replaced in particular uses and the larger groups of the natural resources to which they belong have become less important in economic growth.

The whole question of the importance for unimportance of natural resources in economic growth to which so much attention has been devoted in the last 10 years, generously financed by some research foundations, appears rather banal. In the relation that one may call the resources function, all relevant variables are important (Ciriacy-Wantrup, 1968). Nobody has yet claimed that natural resources are irrelevant in economic growth. In the continuously changing interplay of challenge and response between natural environment and human culture, it is meaningless to say that the challenge is less important than the response. Both are a part of the same system—that of human ecology. If there were no challenge, there would be no response. And, I may add, challenges may yet appear that will prove overwhelming to the response.

Role of Economic Institutions in Economic Growth : Factors or Decision Systems?

The remainder of my allotted time will be devoted to a more meaningful subject—the role of economic institutions in economic growth. In this connection, the mathematical models of economic growth need not be considered because, as stated above, economic institutions are not a part of their input. The output from a mathematical-statistical formulation depends on its input. If one is interested in explaining economic growth as a historical phenomenon and in development policy in the political reality of today, such models have little to offer. Good critiques by economists concerned with development policy are already available.[3] Rather, I should like to focus on the treatment of economic institutions in the book on traditional agriculture mentioned earlier.

To avoid misunderstanding, the main theses of the book are not under discussion. Within the assumptions and terminology employed, traditional agriculture can be regarded as efficient; and its marginal productivity of labour is greater than zero. Neither is it my main criticism that technological change is segmented into "new" factors in order to speak of "shifts" of a global production function—although understanding is scarcely advanced by such terminology.[4] Historically, technological change has proceeded through packages of interrelated changes of many factors. The nature of these interrelations over time, the *gestalt* of technological change, if you like, is the phenomenon that needs understanding rather than the appearance as a deus ex machina of quantities of "new" individual factors. I mention this merely because institutions are subjected to the same segmentations as technological change. My main criticism is that the book presents a conceptually insufficient model of economic growth because the role of institutions is left out.

Sometimes the author himself seems vaguely aware of this insufficiency. He is "puzzled" by economic decisions under a feudal land-tenure system and "baffled" by the economic behaviour of plantation owners under a system of slavery.[5] Still, none of the economic institutions that are significant for resource use—such as the systems of owning and using property, the systems of water rights, taxation, collective and cooperative organisations, and quasi-governmental agencies like public districts—is mentioned, much less systematically treated. Economic growth proceeds in an institutional vacuum as far as natural resources are concerned.

In this vacuum filled by the belated admission of institutions as factors (Schultz, 1968) the services of which are supplied in kind and quantity in accordance with the demand for them until a demand-supply equilibrium is reached.[6] This comforting notion is presented as a "theory" in a field in which it is claimed that "there are virtually no terms of reference, concepts with specifications that can be identified, and no economic theory to guide the analysis" (Schultz, 1968). The facts are quite to the contrary. Institutions have been the central focus of the study of social organisation for more than a century, both by adherents of the theory of economic determinism of institutions and by its opponents.

While Marx. Engels and Kautsky emphasised the role of technological change in economic growth and the economic determinism of institutions, they, as well as non-Marxian adherents of economic determinism, were too aware of the relations between economics and social organisation to press institutions into a simplistic demand-supply scheme.[7] They recognised that the demand for changes of institutions is always opposed by demand for the status quo or demand for change in the opposite direction. They recognised also that a new equilibrium between conflicting demands on the one side and institutional change on the other is a long and hard struggle and is brought to conclusion by political rather than economic forces and tools.

The position of scholars who do not accept the theory of economic determinism of institutions—like Emile Durkheim in France; Max Weber, Werner Sombart and Eduard Hahn in Germany; and W.I. Thomas and Talcott Parsons in the United States—is not affected by this recent oversimplified version of economic determinism (Durkheim, 1893; Khan, 1911; Parsons, 1937; Sombart, 1928; Thomas, 1918-20; Webber, 1927, 1930).

What, then, is the essence of economic institutions? We may conceptualise an institution as a social decision system that provides decision rules for adjusting and accommodating, overtime, *conflicting* demands (using the word in its more general sense) from different interest groups in a society. A change in the demand from one interest group, therefore, rarely effects a change in institutions. A feudal land-tenure system does not change merely because serfs or tenants demand it. A water-rights system based on the riparian doctrine does not change simply because non-riparians demand it. A system of taxation does not change because one or even the majority of taxpayers demand it. It all depends on the effectiveness of different demands, that is, on the relative weight of the interest groups from which demands originate. The locus where conflicting demands meet and where the relative weight of interest groups is determined is the political arena rather than the marketplace. For this reason, changes in economic institutions are usually slow and often require political changes and sometimes outright revolution. Changes in the feudal land-tenure system in Europe took centuries. In South America the change of similar land-tenure systems is still in progress. A hundred years after abolition, the economic effects of slavery in the United States are still present. It took half a century and required a constitutional amendment for the appropriation doctrine to obtain co-equal status with the riparian doctrine in the water-rights system of California. Tax reform in the United States is talked about for decades; but, as every newspaper reader knows, accomplishments are slow.

Demands for institutional change may not even originate from those groups directly involved, such as the landlords and tenants in tenure systems. The demand may come from other groups because of what one might call the externalities of land tenure—that is, the benefits and costs incident on social groups other than landlords and tenants. This is frequently true for the collectivisation of agriculture in order to promote industrial growth and the stability of a socialist political system. Other examples are tenure changes in fugitive resources—such as fisheries, range, oil and gas—to further the social objectives of conservation. We are presently studying a case of this kind for a nomadic group of tribes in East Africa—the Masai. Here, tenure changes are demanded by the government in order to conserve the range resource and the game herds sharing use of the resource with the Masai livestock. The game herds, in turn, are important for economic growth as the basis for a flourishing tourist trade and large foreign-exchange earnings. If the demand for institutional change originates outside the groups directly involved, the changes required are even more clearly political and sociological and usually involve various degrees of "persuasion".

Economic institutions conceptualised as social decision systems provide decision rules both for the use of resources and for the distribution of the income stream derived from such use. This distribution has strong effects on the demand for institutional change. But such income effects are frequently just the opposite of those claimed by the doctrine under discussion. In other words, the demand for institutional change is increased if the income stream of a social group is decreased absolutely or relatively. The demand for changes in the feudal system in Central Europe was increased through dispossession of the peasants in the course of enclosures, and the demand for labour legislation in England was sparked by the misery of the working class during the first stages of the industrial revolution.

Our conceptual framework is not complete by designating economic institutions as social decision systems. I have tried to show elsewhere that economic institutions operate on the second level of a three-level hierarchy of decision systems (Ciriacy-Wantrup, 1967). An understanding of the relations between the three levels is necessary for devising public policies. On each level, the structure, the functioning and the performance of the decision system can be studied both conceptually and observationally. For economic institutions affecting private conservation decisions, that is, decisions on the first level, such a study has long proved its usefulness (Ciriacy-Wantrup, 1968). Water institutions likewise have been analysed within this framework (Ciriacy-Wantrup, 1956, 1967).

To bring this framework closer to the realities of development policy, I should like to illustrate it by some observations regarding the role of water institutions in northern India under the impact of the green revolution—the rapid spread of a new cereal technology based on varieties with greater capacity to combine with complementary inputs, such as labour, fertiliser, and especially water. Because of time limits and because our study is not completed, I can only sketch the main issues.

Role of Water Institutions in Northern India under the Impact of the Green Revolution

The results of the Green Revolution, expressed in yields per acre, increase in acreage of high-yielding varieties and total production need not be recounted here. This revolution has spread so rapidly in northern India, Mexico and several other countries because it requires the least institutional change of any new technology that I know of, at least in the initial stages. In subsequent stages, however, institutional changes will become necessary. Of these, changes in water institutions will be the most essential and the most difficult to bring about.

My concern is that the doctrine discussed in the preceding sections will lead to complacency regarding the institutional difficulties that lie ahead in realising the full benefits of the green revolution and avoiding serious setbacks. According to this doctrine, there are no such difficulties because farmers themselves will accomplish those changes of water institutions that become necessary.[8] Let use take a closer look at this projection.

In northern India, irrigation has been based for centuries on groundwater lifted from open wells by the Persian wheel, a simple but ingenious device to develop groundwater resources where they are available in large quantities not more than some 40 feet below surface. This is the situation in large parts of the extensive Indo-Gangetic Plain.

Groundwater use is usually possible on a more individualistic basis than surface water use. This is especially true for the relatively small, low-capacity development of groundwater based on the Persian wheel. The farmer can build, maintain and operate the water facility himself or with the help of a small number of villagers. Water is available to the individual farmer where he wants it, when he wants it, and in quantities and over time periods that are under his own control.

An irrigation economy dominated by the Persian wheel differs greatly from one that is based on diversion from large rivers, such as in ancient Egypt and Mesopotamia, or in modern times, from large multipurpose reservoirs. Here, the individual farmer is dependent on other farmers for building, maintaining and operating the diversion facilities and the canal system that distributes the water. He must take his water at certain points on the canal. He must take his turn as to when the water is available to him. Quantities of water are rationed to him, and these rations may vary over time for reasons outside his own control. What is the effect of these fundamental differences in the irrigation economy on the development of water institutions?

In a diversion economy, water institutions—such as water rights, water districts, a water master, and public agencies engaged in building, maintaining and operating facilities—are a necessary condition from the beginning. In a water economy based on the Persian wheel, such institutions are not needed and do not develop. Groundwater institutions become a necessity only when water use by one farmer affects his neighbour. This happens when the Persian wheel is replaced by modern deep-well pumps. Here the neighbour is affected by the pumping cone and by the seasonal and often secular depletion of the resource because of high-capacity pumps.

In countries where an irrigation economy based on diversion preceded the need for modern groundwater institutions, these institutions could be built on the experience and mental attitudes already formed. These relations between groundwater institutions and surface water institutions I have traced elsewhere (Ciriacy-Wantrup, 1955). But such experience and such attitudes are not available in northern India. There, water institutions must be created from scratch.

The Green Revolution has created a need for more water. Groundwater tables are beginning to fall and will force the replacement of the Persian wheel by deep-well pumps. Increasing size of farms, related to the Green Revolution, will operate in the same direction because of the Persian wheel's relatively low capacity and high labour requirements. Increased irrigation will increase the need for drainage. Water allocation problems between users, between uses, between States and regions, between areas of origin and areas of destination, and between State and federal jurisdictions loom on the horizon.

To solve these problems, groundwater laws are needed to regulate the number and spacing of wells, the quantities pumped per well, the periods of pumping, and the provision and operation of recharge facilities. This will require adjudication of water rights and integration of groundwater and surface water development and, thus, an integrated water rights and management system.

As a key element of such a management system, the water district has proved its usefulness in the United States and elsewhere. It is flexible with respect to its role as a part of local and regional government. It is flexible also with respect to the participation of social groups outside of agriculture that need additional water supplies. Water districts, possibly with greater governmental participation and supervision than in this country, are needed.

Water institutions of this kind will develop over time. But the individualism of the farmer, the local community, and the region, shaped over centuries in an irrigation economy dominated by the Persian wheel, requires a special effort by researchers and policy-makers to create and to make acceptable the water institutions that are needed. Complacency based on a simplistic doctrine of the role of institutions in economic growth is not at all warranted and will become dangerous if effort in research and in public policy is thereby deflected from this important area. What India's agriculture needs most at the present juncture are competent economists, fully aware of the intricacies and the difficulties in the development of water institutions. Otherwise, the cornucopia of the Green Revolution may indeed prove Pandora's box, to paraphrase the title of a perceptive recent paper (Wharton, 1969). A laissezfaire attitude toward institutions is no less inappropriate than regarding them as constraints.

NOTES

1. According to Schultz, 1964 this influence operates solely through profitability of tenant farming. Profitability is regarded throughout as a "strong explanatory variable". Edmundo Flores' emphasis on land reform as the basic issue in increasing Mexico's agricultural productivity is explicitly rejected.
2. Because of these weights, the study by Goldsmith, Brady and Mendershausen (1956) shows a decreasing share of *all* land in the wealth of the United States, but this study applies only to the period 1896-1956 and is subject to a number of statistical difficulties with respect to comprehensiveness and valuation.
3. See the discussion of the Ranis and Fei and Jorgenson models by John W. Mellor (1967). Also, see the discussion of institutional influences by John M. Brewster (1967).
4. Since multi-dimensional relations are involved, one must, strictly, speak of movements to production hypersurfaces.
5. "Why many of the farmers who own and are responsible for the operation of very large farms, especially in some parts of South America, do not successfully engage in the search for modern agricultural factors is a puzzle." "Why better health and longer life were not realised (for slaves in the antebellum South) is baffling." (Schultz, 1964)
6. The author must be aware that the economic definitions of demand, supply and equilibrium do not apply in the cases he discusses. Clarity is not served by interchanging the more general sense of a word with its precise scientific connotation.
7. Compared with Marx and Engels, Kautsky has not gained a wide reputation in this country. He was, however, an economic theorist of some stature, who was especially interested in the role of agriculture in economic growth (Kautsky, 1902, 1919). Besides Marx. Engels and Kautsky, a long list of non-Marxian works could be mentioned. The best known in the United States is probably Charles A. Beard [1].
8. As Professor Schultz puts it, farmers "will join with neighbours to acquire tube-wells and to undertake minor investments to improve the supply of water. Both tenants and landowners will also use whatever political influence they have to induce the government to provide more and better large-scale irrigation facilities".[22, p. 1118]

REFERENCES

Beard, C. A., 1918 : *An Economic Interpretation of the Constitution of the United States,* New York, The Macmillan Company.

Brewster, J. M., 1967 : "Traditional Social Structures as Barriers to Change," *Agricultural Development and Economic Growth,* Ithaca, New York, Cornell University Press.

Ciriacy-Wantrup, S. V., 1936 : *Agrarkrisen and Stockungspannen zur Frage der langen 'Welle' in der Wirtschaftlichen Entwicklung,* Berlin, Paul Parey.

———, 1956 : "Concepts Used as Economic Criteria for a System of Water Rights," *Land Econ,* 32 : pp. 295-312, November.

————, 1968 : *Resource Conservation : Economics and Policies*, 3rd ed., Berkeley University of California Division of Agricultural Sciences.

————, 1955 : "Some Economic Issues in Water Rights," *J. Farm Econ.*, 37, pp. 875-885, Dec.

————, 1964 : "The New Competition for Land and some Implications for Public Policy," *Natural Resources J.* 4, pp. 252-267.

————, : "Water Economics : Relations to Law and Policy,: in *Waters and Water Rights : Eastern, Western Federal*, ed. Robert Emmet Clark, Indianapolis, The Allen Smith Company, 1967, Vol. I, pp. 97-431.

————, 1967 : "Water Policy and Economic Optimising : Some Conceptual Problems in Water Research," *Am. Econ. Rev.* 57, pp. 179-189, May.

Durkheim, Emile, 1893 : *De'a Division do Travail Social; Etude sur Porganisation des societes superieures,* Paris, F. Alcan.

Fei, John C. H., and Gustav Ranis, 1964 : *Development of the Labour Surplus Economy: Theory and Policy,* A publication of the Economic Growth Centre, Yale University, Homewood, Illinois, Richard D. Irwin.

Goldsmith, Raymond W., Dorothy S. Brady, and Horst Mendershausen, 1956 : *A Study of Savings, in the United States*, Vol. 3, Princeton, Princeton University Press.

Hahn, Eduard, 1911 : *Die Entstehung der Pflugkultur,* Heidelberg, Carl Winter's Universitatsbuchhabdlung.

Jorgenson, Dale W., 1961 : "The Development of a Dual Economy," *Econ J.*, 71, pp. 309-334, June.

Kautsky, Karl, 1902 : *Die Agrarfrage: Eine Uebersicht ueber die Tendenzen der Modernen Landwirtschaft, und die Agrarpolitik der Sozioldemokratie, Stuttgart,* I.H.W. Dietz Nachf., Inc.

————, 1919 : *Die Sozializierung der Landwirtschaft,* Berlin, Paul Cassirer.

Mellor, John W., 1967 : "Toward a Theory of Agricultural Development," *Agricultural Development and Economic Growth*, Ithaca, New York, Cornell University Press.

Parsons, Talcott : *The Structure of Social Action; A Study in Social Theory with Special Reference to a Group of Recent European Writers*, New York, McGraw-Hill, 1937.

Potter, Neal and Francis T. Christy, Jr. 1961 : "Employment and output in the Natural Resource Industries, 1870-1955," in *Output, Input and Productivity Measurement,* Studies in Income and Wealth, a report of the National Bureau of Economic Research, Princeton University Press, Vol. 25, pp. 109-145.

Ranis, Gustav and John C.H. Fei, 1961 : "A Theory of Economic Development," *Am. Econ. Rev.*, 51, pp. 533-565, Sept.

Schultz, Theodore W., 1960 : "Connections Between Natural Resources and Economic Growth," in *Natural Resources and Economic Growth,* Papers presented at a conference held at Ann Arbor, Michigan, April 7-9, 1960, ed. Joseph J. Spengler, Washington D.C., Resources for the Future, Inc., 1960, pp. 1-9. (Similar views are expressed in papers by Chandler Morse, Herold J. Barnett, John H. Adler, and others).

————, 1968 : "Institutions and the Rising Economic Value of Man," *Am. J. Agr. Econ.*, 50, pp. 1113-1122, Dec.

————, 1960 : "Land in Economic Growth," in *Modern Land Policy*, ed. J.B. Halcrow, Urbana, University of Illinois Press, pp. 17-39.

————, 1964 : *Transforming Traditional Agriculture*, New Haven, Yale University Press.

Som Bart, Werner, 1928 : *Der moderene Kapitalismus* 3 Vols. in 6, Munchen and Liepzig, Duncker und Humblot.

Thomas, W. I., 1918-1920 : *The Polish Peasant in Europe and America : Monograph of an immigrant group,* 5 Vols., Chicago, University of Chicago Press.

Weber, Max, 1927 : "The Evolution of the Capitalistic Spirit," in *General Economic History*, trans, Frank H. Knight, Glencoe, Illinois, The Free Press, pp. 352-369.

———, 1930 : *The Protestant Ethic and the Spirit of Capitalism*, trans. Talcott Parsons, London, George Allen and Unwin, Ltd.

Wharton, Clifton R., Jr., 1969 : "The Green Revolution : Cornucopia or Pandora's Box?" *Foreign Affairs*, 47, pp. 464-476, April.

4

LOCAL KNOWLEDGE AND SUSTAINABLE AGRICULTURAL DEVELOPMENT IN INDIA

BRIAN J. MURTON

OF LATE, the term "sustainable agriculture" has permeated the agendas of international agricultural research organisations. Sustainability implies balancing agricultural development and environmental preservation in a culturally and socially acceptable manner. However, despite the importance and popularity of the sustainability theme, there is still little understanding of just what constitutes a sustainable as opposed to an unsustainable development path. Clark (1987) has pointed out that progress in implementing sustainable development policies has lagged and that scholars have been slow to examine critically the basic assumptions and practically trade-offs that lie at the heart of would-be sustainable resource management.

One of the dimensions emphasised in the growing literature on sustainable development in agriculture is the importance of local control and participation in development planning, and especially for a small-farmer orientated set of agricultural policies. This approach involves a research and development agenda for resources-poor farmers which would be highly specific to local ecological conditions, would be founded on a thorough knowledge of local agricultural practice, and would be participatory in style. It is already claimed that these has gone far in Africa (Richards, 1985; 1886), and there is growing evidence that this approach is being incorporated into agricultural development strategies in Southeast Asia (Marten, 1986; Charoenwatana and Rambo, 1988).

It is not that the various elements involved in this approach are new. Indeed, many of them have received considerable scholarly attention for a long time. In the remainder of this paper examples from India are used to stress the role that traditional practices and knowledge can play in the agricultural development process. The first section of the paper overviews the ways in which Indian farmers traditionally have managed soil, water, climate, slope and space. The second section reviews indigenous knowledge about the environment of agriculture. The last section raises questions about the adequacies of a populist approach and suggests ways of resolving the difficulties involved in accomplishing such a task.

Traditional Resource Management Techniques and Practices

The various ecological and technological strategies which continue to survive in India were essential mechanisms in maintaining long-term stability in the traditional resource complexes of the region. To survey the vast array of traditional technologies, techniques and practices would require more space than is available here (Murton, 1980). Rather, for purposes of explication attention is directed to five major resource categories and how there are managed: soil; slope; water; micro-climate; and space (Wilken, 1987).

Soil Management

An impressive array of soil management techniques is employed by Indian farmers. These techniques can be divided into three major groupings: soil fertility practices; tillage practices; and field forms. Soil fertility is maintained or improved by a variety of animal, green, and human manures, usually produced within the farm area. Cattle dung remains the preferred organic manure, but in many places a shortage of firewood means that it is more often than not used as a fuel for cooking.

A striking feature of soil fertility maintenance and enhancement practices is that most of them are only applied to irrigated land. Only the penning of sheep and goats, the use of ash (shifting cultivation), fallowing, crop mixing, the use of leguminous crops, and crop rotation are found on rainfed land. Furthermore, much of the manure of one type or another, in millet-growing areas, as well as in wheat and rice staple areas, is applied to land destined for cash crops such as sugarcane, vegetables, tobacco, and betel, rather than for staple food grains.

Various tillage practices affect tilth for better or for worse. Most land receives multiple ploughing, usually at intervals of a week or two. Although ploughings are shallow this does result in extremely fine tilth in the top four to eight inches. Hoes are used throughout the region to break up clods left after ploughing. Harrowing using a variety of harrow-type implements is widespread, but it is especially important in the black soil parts of peninsular India, where it often supplants ploughing. Clod-breaking by hoeing, by pulling beams across the fields, or by using hammers or mallets, is universal, as is puddling in rice-grown areas. The heating of the top four to eight inches of the soil which kills insect pects is a by-product of burning in shifting cultivation areas. Certain kinds of field forms are important soil-management practices in Kerala, West Bengal, Bangladesh, and Orissa.

Slope Management

Slope management is closely related to soil management and water management, especially where erosion control and irrigation are practiced. Terracing is, perhaps, the most widespread and best-known form of slope control in India. It is possible to distinguish many different types of terraces, and in India terrace forms fall into the following categories: silt-trap dams; crude stone walls across sloping surfaces; stone-walled terraces for both dry and irrigated crops; series of irrigated fields below water sources. Field levelling is a widespread slope management technique found in association with terracing. It is especially important in irrigated areas where it is an aspect of water control as well. Planted vegetation, usually hedges, is in use in several parts of India where terrain is either rolling or sloping, often in association with low piled rock walls, which creates very distinctive, almost "bocage" type landscapes.

Water Management

Management of water resources includes some of the oldest and most widespread of traditional practices. This stems in part from the fluid nature of water which permits the use of gravity to perform such work functions as collection, transfer, and application. Most Indian systems simply interrupt and reorganise the overland portion of the hydrologic cycle to secure a supply of irrigation water. Water storage is quite complex as it involves artificial devices (tanks) for true water storage, rather than simply diversion. Water from wells, rivers, and tanks is raised or transferred by a variety of lifting devices, and water is diverted from streams through the construction of embankments in their beds. Water delivery systems almost universally depend upon gravity flow through open canals. Once at the fields most water is applied through surface flooding, although where water is scarce it is applied by hand from pots. Various forms of bunding controls the flow of water through fields.

Numerous devices are used to measure the allocation of water: graduated stakes; graduated sundials; the position of moon and stars; pots from which water drops through a hole. It is usual to have a designated official supervise the actual apportioning of the water, although in some areas landholders construct, maintain, and control, water allocation to tenants. Elsewhere, farmers usually build, maintain, and repair small-scale facilities themselves, but large works which were formerly constructed and repaired with state or religious institutional support, are now maintained by government departments. Medium-sized facilities — most often large tanks, diversion devices, and less often wells — are built and maintained by one or several village communities, or by the government.

Drainage systems also exist: downslope systems in terraced ricefields; water lifts; and ditch systems. Flood prevention systems — levees along rivers and embankments and dams across tidal estuaries to keep out salt water — are found in low-lying delta areas of the major river systems.

Microclimatic Management

Traditional farmers in India are not able to cope with enormous energy fluxes and mass transfers involved in macro- or meso-scale climates but they have achieved considerable control over individual plant or small plot microclimates. Perhaps the most basic way of modifying microclimates is the management of the amount of sunlight entering the soil-plant system. Shade management is widespread: shade reduction through forest clearance; plant spacing; shade enhancement; mixed cropping. The intensity of solar radiation received at the surface is also changed through the use of raised ridges and walls. Heat and moisture fluxes are controlled by tillage practices and, to a much more limited extent, by mulching. Tillage practices serve the multiple purposes of increasing precipitation absorption, reducing heat flows between surface and subsoil, controlling weeds, and improving root and tuber development. Tillage methods produce a fine surface tilth, a "dry soil mulch". The harrowing and hand pulling of weeds reduces transpiration losses. Windbreaks of one type or another are also used in several parts of the region.

Space Management

Although Indian farmers live and work in a relatively small world, distances and spatial arrangements enter into their decisions. India has had numerous large cities and smaller towns with permanent markets for nearly three millenia, and in addition there is a well-developed network of fairs and periodic markets. Within local regions virtually every village has access to a periodic market, and commonly a fair at an important temple or shrine.

Distances and distribution of fields in relation to farmstead is also important. Location is an attribute of land at the village level, just as is soil quality and water availability. This is reflected in decreasing landuse intensity from the settlement site. Another aspect of farmstead-field distance is the fragmentation of land holdings. Farmers prefer to have fields located in all zones around a settlement site, so they can share in all types of production, thus minimising the risk of total crop failure. Space also is managed at the plot level, where mixed cropping, which includes multistoried cropping, enables maximum use to be made of both horizontal and vertical space.

Systems of Agricultural Knowledge

These specific agricultural techniques, practices, and technologies, however, are but the surface manifestations of an underlying system of knowledge. Such knowledge is basic to farmer's management strategies yet it is often overlooked. This information is extremely varied, ranging from detailed knowledge about specific plants to broader patterns of categorisation that tell much about how a particular group views the world it inhabits. Broadly speaking, such indigenous knowledge falls into two categories (Brosius, Lovelace and Marten, 1986). On the one hand, there is knowledge that is inherited from generation to generation and is generally shared by most members of society. The second category includes information gained through individual experience — empirical observations made by individual farmers during the course of agricultural activities.

No systematic accountings of either type of knowledge have been made for India, but relevant material has been described in a number of places (for example, Balasubramanian and Gopalkrishnan, 1953; Grierson, 1885; Randhawa Sivaraman and Naidu, 1961), and reviews of cosmological, spatial, and temporal concepts that make nature intelligible are beginning to appear (Zimmerman, 1987; Murton, 1981; 1987). The theme is not without its detractors. Equally convincing evidence can be presented for the ecological rationale of traditional methods and even for the reverse argument that cosmological concepts and kinship relationships are best illuminated by the understanding of physical processes and their management. Nonetheless, indigenous knowledge concerning the general environment of agriculture is considerable. In this section three questions are used to organise this material: (1) what are the principal categories of resource discrimination and what are the bases upon which they are made; (2) how are agricultural activities coordinated in cycles of use and access that regulate the timing of the system; and (3) how are limited and unpredictable resources combined in such a manner to make them less limited and more predictable.

Categories of Resource Discrimination

In a previous section five major categories of resource discrimination (soil, slope, water, micro-climate, space) were discussed from the perspective of management practices and techniques. In addition, Indian farmers have terminological domains which include time, plants and animals, and all of these are further elaborated in taxonomies. Indian farmers have classifications of landscape morphology, land types, soils, vegetation, crops, and of macro- and micro-climatic resources.

Agricultural classification reflect local characteristics of the biophysical environment. Throughout the region land is initially classified as either "wet" or "dry". This distinction between irrigated and rainfed land is of major importance because distinctive types of management and conservation practices are related to wet and dry land. In the Ganga Valley and delta areas land is further classified in terms of elevation above water level. Soils are most often initially classified according to their colour (e.g. "red" or "black"), and then in terms of very fine textural discriminations, as well as distance from settlement sites. Crop varieties are classified in terms of obvious seed colour, height, seedhead shape, duration, taste, and other visible characteristics.

Coordination of Agricultural Activities

The coordination of agricultural activities has both temporal and spatial aspects. Traditional systems of dividing time are common to all parts of India. Three methods of computing yearly cycles are in use (solar, lunar, fasli or revenue year), but in terms of agriculture the solar year which begins in mid-April is the most important as its beginning more or less coincides with the thunder showers of April which herald the coming monsoon, the beginning of the agricultural season.

The exact number and dates of seasons in the year varies from place to place, but there are usually six, of two months each. As important as recognised seasons, are sowing periods which are highly localised, reflecting appraisals of the moisture environment in relationship to particular crops. But perhaps the most important devices for regulating the timing of agricultural operations are astral periods or asterisms (*nukshatra*). These are formulated in terms of the relationships of the sun, the twelve signs of the zodiac and twenty-seven stars. There are twenty-seven periods, to each of which are assigned agricultural operations and tasks.

If time is organised into years, months, seasons, and astral periods, space is organised into a number of land soil-crop associations. Throughout India certain crops or crop associations are related to certain land and soil types. These interrelationships are models which farmers use to guide them in their agricultural activities: indeed, they are culturally prescribed ways of doing things.

However, Indian agricultural knowledge is not confined to a sum of unconnected parts. Environmental appraisal, as well as agriculture, requires that terms be linked in webs of associations: sorghum intersects with certain land types, soil/water conditions, insect pests; rice with others; wheat with yet others. Different seasons intersect with the availability of different forage plants, different weather conditions intersect with different diseases, and different terrains with different plant associations. Thousands of these simple linkages constitute the building blocks of traditional Indian agricultural strategy: the ways in which resources are combined and timed.

Resource Predictability

An important aspect of fold knowledge concerning the environment of agriculture is how resource-use activity is coordinated and timed, and how limited and unpredictable resources are made less limited and more predictable. At an abstract level, time and space are organised into sequences and segments, linked into weeks of association. A wealth of knowledge related to these serve as a guide to agricultural operations. This accumulation of experience is formalised in the form of folk songs, sayings, and proverbs (Gover, 1871; Benson, 1933). The months, the year, the seasons, and especially the asterisms, all have associated with them a vast array of proverbs about every aspect of tillage and agriculture operations. Proverbs in general are a collective product of many minds in a cultural group. Agricultural proverbs in particular can be regarded as a reflection of a group's accumulated experience in an environment. Traditionally, they were transmitted orally and were normally learned by small boys as they helped their fathers in the fields. In Madras Presidency, for example, agricultural proverbs were collected and published by the Department of Agriculture. About 20 per cent of these proverbs deal with some aspect of climate and weather, especially seasonal cycles, types of weather expected at different times of the year, problems resulting from deviations from normal weather, and the ways in which an uncertain environment could be made predictable and knowable. Basically such proverbs deal with resource predictability. They represent empirical generalisations of experience in the South Indian environment, and can be regarded as statements of what cultivators expect to occur. For example, rain in the month of *cittirai* (mid-April to mid-May) is viewed as undesirable as it is too early for cultivation to begin. At this time throughout Tamilnadu, rainfall increases, and then usually declines between mid-May and mid-June (*vaikaci*). Throughout most of Tamilnadu the southwest monsoon rains do not begin on a large scale until mid-June to mid-July (*ani*), and the importance of adequate rain in this period is reflected in a number of proverbs. Furthermore, it is believed that if *ani* is dry, *ati* and *avani* (mid-July to mid-September), also are likely to be dry. Other proverbs point out the uncertainty of rain in *ati*. The peaking of rains associated with the northeast monsoon are reflected in numerous proverbs referring to *purattaci* (mid-September to mid-October) and *aippaci* (mid-October to mid-November). However, the uncertainty of rain, even at this time of year, also is acknowledged.

After mid-November, the decrease in rainfall is reflected in the proverbs. The variable weather of *markali* (mid-December to mid-January) is noted as is the beginning of the dry season at the start of *tai* (mid-January to mid-February).

Agricultural tasks are tied to months and weather conditions in many proverbs. The importance of ploughing in certain months reflects an accurate appraisal of the weather. Seed time is also related to months; inappropriate times to sow are noted.

Furthermore, the proverbs describe the uncertainty of rainfall. Tamil peasants have made assessments of the chances of rain at different times, exemplified by the use of the word "if" in the proverbs. It is also possible to distinguish attitudes toward uncertainty. A first is to make it determinate and knowable. In this approach the essential randomness that characterises an uncertain pattern is replaced by a determinate order in which a sequence of events can be predicted. Another method of eliminating uncertainty is the art of "wishing it away" by denigrating the quality of a natural event to the level of the commonplace, or conversely, of elevating it to a unique position and ascribing its occurrence to a freak combination of circumstances incapable of repetition.

Many proverbs try to make rainfall uncertainty determinate and knowable. Several devices are used to do this. All are concerned with predicting future occurrences of rainfall, or the lack of it, and the consequences of one or the other state. Furthermore, all are connected with natural phenomena or with astrological observations. The largest number of proverbs are those concerned with rain in months and astral periods.

The next largest set of proverbs that introduce determinacy deal with forecasts based on the occurrence of thunder and lightning. Lightning is generally held to foretell rain; thunder is deceptive. Thunder during the hot season may indicate rain, but later in the season heralds rain. The position of the planets, sun, moon, and stars are also widely used as predictive devices. Wind direction, often in association with a month or asterism, is also widely noted to bring an element of certainty into the uncertain world. Clouds, halos around the sun, rainbows, and red sky are also cited to reduce uncertainty, and the appearance of certain insects are taken as an indication of rain. For example, beetles, butterflies, white ants, dragonflies and mosquitoes are forerunners of rain. Animals and trees are mentioned in a similar fashion.

Local Knowledge, Populist Approaches and Sustainable Development

There is little question that a vast amount of information about traditional management practices and technologies, as well as systems of knowledge about the environment of agriculture exists for various parts of India. The material presented in this paper has demonstrated that shared cultural knowledge about agriculture can be systematically described. However, the incredible diversity of local ecological conditions, local agricultural practices, and local socio-political conditions have barely been alluded to. A truly comprehensive accounting of such richness and diversity is an almost impossible undertaking, other than on a very small scale. Furthermore, no information has been presented on the knowledge that individual farmers gain through experience, and on how this knowledge is generated. Very little is known about the role of experience, both as a source of new information and as a referent and strong reinforcement for received information.

There is also little question that knowledge about agriculture varies among individuals. It also tends to vary with respect to particular groups of people within the larger social unit (age, sex, class, caste, occupation). Such potential variations in traditional knowledge raise many topical as well as methodological questions. For example, how exactly is an individual's knowledge related to the knowledge of the society? Assuming that the new knowledge is ultimately created by individuals, how is social knowledge created or acquired and what factors affect its creation or acquisition? In terms of methodological questions, on the one hand, perhaps the most crucial issue is whether the existence of individual variation in knowledge makes it virtually impossible to isolate knowledge that is truly shared, especially where interviewing of individuals is the primary method of gathering information on the traditional pattern of knowledge.

Obviously, sustainable agriculture development requires on-the-spot attention, involving local knowledge, continuous monitoring, and sometimes vigorous action. Surely then, bureaucrats, policy-makers, and research workers should listen and learn from farmers. The evidence provides a sound basis for supporting a "bottom-up", locally specific populist approach to agricultural development. However, local knowledge of appropriate methods may not be enough. In many cases, the provision of resources and incentives to make development possible for poor farmers requires considerable state involvement. Moreover, most local management practices and knowledge have been evolved in pre-capitalist and/or non-market situations. Today, the market penetrates virtually all landuse in the world, and market opportunities are a strong additional force which puts pressure upon resources. In many cases too, locally enforced authority has broken down and the state is forced to step in to protect resources.

In India, the question of adapting indigenous agricultural technology, management practices, and knowledge to the needs of modern market agriculture is complex and it involves contradictions which are irreconcilable in the short run: mechanisation and yield maximisation versus work spreading and risk minimisation; short-term versus long-term goals; tenurial systems which increase access to land versus consolidation and greater inequalities. The importance of achieving means of sustainable agricultural development which are within local competence is not however, diminished by these difficulties, and ultimately a participatory, local-based approach, may be the only way to make small, resource-poor farmers aware of the potentials of their own resources and traditional management system.

REFERENCES

Balasubramanian, C. and Gopalkrishnan, R., 1953 : "Agricultural Folklore in Malabar", *The Madras Agricultural Journal*, Vol. 40, pp. 55-59.

Benson, C. A., 1933 : "Tamil Sayings and Proverbs on Agriculture", *Bulletin*, Department of Agriculture, Madras, India, No. 29, New Series (Reprint), pp. 1-75.

Brosius, J. P., Lovelace, G. W., and Marten, G. G., 1986 : "Ethnoecology: An Approach to Understanding Traditional Agricultural Knowledge", in Marten, G.G. (ed.), *Traditional Agriculture in Southeast Asia: A Human Ecology Perspective* (Boulder, Colorado: Westview Press), pp. 187-198.

Charoenwatana, T. and Rambo, A. T., 1988 : *Sustainable Rural Development in Asia*, Selected paper from the Fourth SUAN Regional Symposium on Agroecosystem Research, Khon Kaen University, Thailand, July 4-7, 1988. Khon Kaen, Thailand: Farming Systems Research Project and Southeast Asian Universities Agroecosystems Network.

Clark, W. C., 1987 "Editorial: Scholarship for Sustainable Development", *Environment*, Vol. 29.

Gover, C. E., 1871 : *The Folk Songs of Southern India* (Madras: Higginbotham and Co.).

Grierson, C. A., 1885 : *Bihar Peasant Life* (Calcutta: Bengal Secretariat Press).

Marten, G. G., 1986 : *Traditional Agriculture in Southeast Asia. A Human Ecology Perspective* (Boulder, Colorado: Westview Press).

Murton, B. J., 1980 : "South Asia", in Klee, G.A. (ed.), *World Systems of Traditional Resource Management* (London: Edward Arnold), pp. 67-100.

———, 1981 : "Traditional Systems of Agricultural Knowledge and Development in India", in Mohammad, N. (ed.), *Perspectives in Agricultural Geography*, Vol. 4. *Regional Dimensions in Agriculture*, New Delhi: Concept Publishing Co., pp. 401-414.

———, 1987 : "Monsoons in Agricultural Proverbs", in Fein, J.S. and Stephens, P.L. (eds.), *Monsoons*, New York: John Wiley and Sons, pp. 77-102.

Randhawa, M. S., Sivaraman, M. S., and Naidu, I. J., 1961 : *Farmers of India, Volume 2* (New Delhi: Indian Council of Agricultural Research), pp. 97-105.

Richards, P., 1985 : *Indigenous Agricultural Revolution: Ecology and Food Production in West Africa* (Boulder, Colorado: Westview Press).

———, 1986 : *Coping with Hunger: Hazard and Experiment in a West African Rice Farming System* (London: Allen and Unwin).

Saint, W. S. and Coward, E. W., 1977 : "Agriculture and Behavioural Science: Emerging Orientations", *Science*, Vol. 197, p. 733.

Wilken, G. C., 1987 : *Good Farmers. Traditional Agricultural Resource Management in Mexico and Central America* (Berkeley: University of California Press).

Zimmermann, F., 1987 : "Monsoon in Traditional Culture", in Fein, J.S. and Stephens, P.L. (eds.), *Monsoons* (New York: John Wiley and Sons), pp. 51-76.

5

VARIATIONS IN FOODGRAIN PRODUCTION AND EFFICIENCY IN CHINA

GREGORY VEECK AND CLIFTON W. PANNELL

SINCE 1979, there have been profound changes in the rural economy throughout much of Jiangsu Province. The rural reforms, in conjunction with the "Opening to the West", have ushered in a period of unprecedented prosperity, particularly in southern Jiangsu (*Sunan*). Incomes have risen, housing space and quality have improved dramatically, and consumer durables production and consumption have skyrocketed. (J.E.Y., 1986, and Lee, 1984). In general, there is an easily observable growing prosperity. The reforms which have been widely credited for these improvements are actually a series of interrelated economic development policies designed to foster greater commercialisation, diversification, and integration of rural areas within larger economic spheres. (Song, 1983) Central to these goals is the transfer of decision making from pre-1979 commune leaders and provincial planners to individual households and re-defined rural collective institutions. (Domes, 1982 and Crook, 1986) Greater freedom to allocate resources has resulted in major shifts in the employment structure throughout rural Jiangsu. With both national and provincial support, rural industry has seized a significant portion of total industrial output in every prefecture in the province. (J.E.Y., 1986) Increased demand for workers by the rapidly growing industrial sector has transformed the basic economies and employment characteristics of most counties. The extent of this transformation varies by location, with the greatest change occurring in the more developed south. In Wuxi and Suzhou Prefectures, off-farm employment accounted for 68 per cent and 62 per cent respectively of total employment in 1986. (J.E.Y., 1986) In some areas northern Jiangsu (*Subei*), this figure is below 5 per cent.

At the present time, the greater commercialisation, industrialisation, and integration of many rural areas represents the genesis of a different kind of place which falls outside of the traditional urban/rural dichotomy. These places are neither completely rural nor urban in characteristics but rather, a hybridisation of both. McGee (1989) has referred to this type of extended mixed rural yet urban place as "Kotadesasi" in other parts of Asia. These areas may develop with or without government support. In Jiangsu, the promotion of these areas has received full government support at all levels. The policy known colloquially as *LiTu Bu LiXiang* (Leave and Land but not the Township) is designed to limit massive migration to urban areas, while fostering economic development. High levels of urban migration are seen as costly and considered to hamper development efforts. (Zweig, 1987) This policy has sparked a tremendous increase in rural industry and employment. The growth of the industrial sector has fostered higher wages, better transportation and certainly greater opportunities and expectations than would be found in traditional rural areas of China. Farm households in these areas do not necessarily abandon their cropland. Rather, the household economy appears to diversify with the one or two family members finding off-farm employment, and some other members managing the household's contracted land. These extended metropolitan or mixed "Kotadesasi" regions are difficult to delimit in Jiangsu, but there can be little doubt that some new type of place is evolving which seems somehow different from the usual urban/rural transitional zones such as "peri-urban zones" or "zones of urban influence". The difference may depend on the greater permanence of these zones in contrast to the more traditional models.

Assuming the growth of what might be called "Kotadesasi" areas in Jiangsu, a different type of household has concurrently emerged. These households, in one sense, are like households before the commune period (prior to 1958) which derived income from crops, husbandry and cottage industries. The present version of these "mixed economy" households are unique in that in addition to husbandry and cottage industries, some family members could also derive income from construction, transportation, commerce or industry.

It is often suggested that rural development in developing countries must occur in conjunction with a shift of the workforce from agriculture to industrial and service jobs, improved worker productivity, and a hastening of the commercialisation of agriculture. The basis for these arguments is a presumed exodus of workers from agricultural production to off-farm employment. In the case of Jiangsu, the newly-minted industrial workers have not given up their land. Rather, households will, if possible, shift the more able workers off-farm, and rely on children and older family members to perform agricultural tasks. Alternatively, crop cultivation is pursued in the evenings after other work is done. As such households have increased their incomes through off-farm employment, the significance of income derived from agriculture declines. Farm households have grown disenchanted with the low profits from crop cultivation and, when possible, prefer to concentrate their labour and capital on more lucrative sidelines or off-farm enterprises. (Oi, 1986 and Cheng, 1987) In short, it is possible that much of the limited cropland in Jiangsu is controlled by households who are not concerned with maximising profits or maintaining efficient production, simply because attention is focused elsewhere. As the "Kotadesasi" regions spread through the economic reforms and government support, less efficient grain production may be an unanticipated side effect. The hypothesis posited and tested in this research is that households which no longer rely on income from agriculture are not as efficient producers as full time farm households. If variations in production efficiency among types of farm households can be identified in Jiangsu, then the transfer of land to more efficient households would promote the more efficient use of scarce land, capital, and agricultural inputs.

To a large extent, these "mixed income" households are products of the reforms and active government promotion of the development of "Kotadesasi" regions. These households are the new inhabitants of these areas, and as these areas spread, so will such households. If some of these households are less efficient producers, it would appear rational to develop policies to transfer land from these households to others which still focus on crop production. It may be just such concerns that have prompted the promotion of "Specialised" Households (*Zhuanye Nonghu*) and other programmes designed to support increased agricultural productivity.

Background and Study Area

Southern Jiangsu is one of China's most economically advanced areas. The lower Changjiang (Yangtze) Basin has about 1100 mm of precipitation annually, a growing season of approximately 230 days, and generally productive red and yellow alluvial paddy and podzol soils. The region developed an intensive rice and fishing economy more than 1000 years ago. With steady production growth over the centuries, improvements in farming led to more intensive cropping systems based mainly on rice, but including vegetables and cash crops such as mulberry/silkworms, tea, and cotton. The long term stability of agricultural production in the area, coupled with its general prosperity, resulted in southern Jiangsu being referred to as "The Land of Rice and Fish" (*Yu Mi Zhi Xiang*). Throughout much of the dynastic period, and prior to the commune period in this century, *Sunan* has had a long tradition of cottage industries which were used by rural households to supplement household income. Prepared teas, silk, embroidery, woven cloth and baskets, and dried and preserved fruits, fish, and fowl were exported from the region and brought considerable wealth and renown to the area. Fei Hsiao tung has noted that traditionally, much of the disposable income of households in many villages in *Sunan* was derived, not from agriculture, but from cottage industries (Fei, 1983).

The multiple income household has precedent in this area, and the recent shift of households increasing off-farm income actually represents a return to the traditional pattern. When the reforms adopted in 1979 were promoted throughout the rural areas of *Sunan*, their rapid success might in part be credited to this earlier tradition of commerce and cottage industries.

Improvements in transportation developed over centuries. The lower Changjiang region has had a low cost, and effective, transportation system for much of its history. This system, composed of rivers and canals, integrated local and regional economies with the larger political control centres and regions of the North China Plain. Superior transport infrastructure and local government sponsorship of the reforms in Sunan supported mercantile traditions and permitted the rapid adoption of the reforms. These factors in conjunction with a surplus-producing agricultural sector provided an environment conducive to rapid industrial and commercial growth.

This kind of spatial integration has continued into contemporary times, and has made this part of China a laboratory of great significance for analysing conditions of economic growth and development under socialism (Weins, 1982). What happens in regions such as Sunan are valuable test areas for the broader Chinese economy, and may indicate the direction of development policies for the entire economy (Rothenberg, 1987). In addition, the reforms were generally implemented in Sunan at an earlier time than in most parts of central China. The transformation of rural areas, and their workforces, which has already occurred in Sunan, may occur in many other areas in the near future.

Originally, Sunan was selected exclusively for this research. Huaiyin Prefecture in northern Jiangsu (Subei) was later selected to provide a contrast to conditions in the surveyed areas of Sunan. Including this area, interviews were conducted in eight counties incorporated in four prefectures. The four prefectures and eight counties included in the survey are depicted in Fig. 5.1. They are: (1) two counties within Nanjing Prefecture north and south of the Changjiang River; (2) two suburban counties of Suzhou Prefecture — one northeast of the city and one near Lake Tai; (3) a suburban county in Nantong Prefecture as well as a more isolated county along the coast; and (4) Two counties of Huaiyin Prefecture in Subei which reflect the generally lower levels of economic development of the north.

The environments of these four study areas are different. The levels of economic development, integration, and commercialisation of these areas also vary and reflect their linkages to urban centres and larger external economies. This variation in linkages affects the current possibilities for marketing surplus commodities produced on the farms. More importantly, they also result in different labour opportunities for peasants seeking off-farm employment, and in that way may strongly affect the magnitude and composition of total farm family income.

Data based for this study were drawn from a larger survey conducted on family farm production and income during late 1986 and throughout 1987. These data include information on household income and income sources, household demographic information, information on land contracted by the household, and complete data on unit-area inputs and outputs for the production of wheat and rice. The data were collected through interviews and an extensive questionnaire completed by each household within the survey. In addition, interviews with agricultural extension personnel and government cadres were most useful, and provided considerable insight to the collected data.

Again, the purpose of the research was to determine if differences in sources of household income have any relationship to variations in foodgrain production efficiency. Production efficiency, in this analysis, was determined by the net return per *yuan* invested. At the time of the analysis one U.S. dollar was equal to 3.71 Chinese *yuan*.

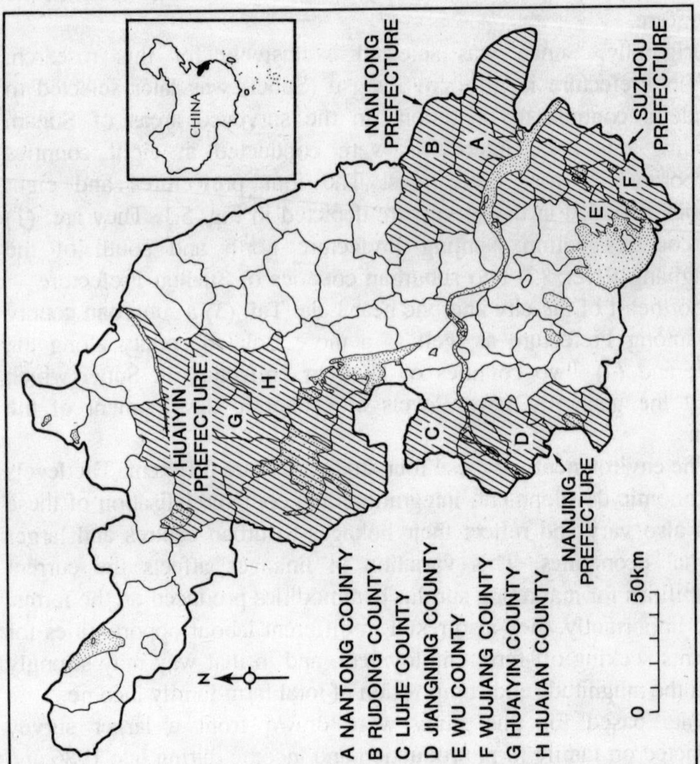

FIG. 5.1: Jiangsu province: political units of the survey.

Previous analysis of these data suggested that households within each statistical unit of the survey (counties) exhibited a notable degree of variation in production efficiency from household to household. That is, the location of each household was a poor estimator of relative efficiency of production or for expected returns on capital investment. Most official data regarding production efficiency is reported at the county level. Household data aggregated to this level masks considerable variation, and county averages do not appear to be suitable for comparisons of production efficiency. While location does not appear to be a good indicator, it was hypothesised that different "types" of households, once identified, are a more salient predictor of production efficiency for foodgrain production.

To identify the posited "types" of households, and to classify the surveyed households into these types, a sorting routine employing Principal Components Analysis and Ward's Minimum Variance Clustering Procedure was performed.

Classification with Principal Components Analysis

Before the households could be classified, the appropriate criteria for distinguishing "household types" were identified. The first step in this procedure was to determine what demographic or income variables explain the variance observed among the sample households. Principal Components Analysis (PCA) is a method for expressing information in an alternative form which is often more succinct than the original (Jackson, 1983). The method is mathematical and does not require a theoretical design or assumptions of causality. It is simply a restatement of a given data set in a new way. PCA is often used for variable reductions, but it can also be used to eliminate multicollinearity in a data set — through the transformation of variables to components which are orthogonal to each of the other components (Ray, 1982). For each component created, each household in the sample will have a component score, which describes the location of that observation on the component.

In a sense, these new variables are denser manifestations of the original variables. Once these component scores have been created through PCA, they can be used like any other variables to create groups through cluster analysis (Ray, 1982). The cluster analysis was used to assign membership in defined groups to each household in the sample. Because the component scores derived from the PCA were used to determine the clusters, the clusters represent the types of households delimited by the PCA.

Once each household was assigned membership in a "household type", production budgets were created for each type. A production budget is a tabular listing of all inputs and the resulting outputs, which can be used to calculate the average net returns for a given unit of land. In this case, the production budgets were designed to show the costs and returns for a single *mu* of land which is planted to the wheat-rice cropping system by each household in the sample. The budgets were developed for each crop, as well as for the annual cropping system. Production budgets are a simple way of calculating the relative effectiveness of capital investment and the net returns of investment for each group of households identified through the principle components/cluster analysis procedure. The format of the budgets follows one developed by Caukins after his evaluation of the appropriate use of western agricultural economic research in China (Caukin's, 1983). After the production budgets were created, multiple comparison Tukey tests, also called the "honesty significant difference test", were performed on each variable incorporated within the production budgets (Zar, 1984). The tests determined if the means of each variable which represent the means for each group of households were significantly different from the variable means of the other groups. If output, net returns and expenditures of labour for each household type prove to be significantly different ($p = .05$), then the production efficiency of wheat and rice for each household type must be considered to be different.

The unit area yields of grain on most of the surveyed households were excellent, with a total sample average of over 11,800 kg/hectare/year. The cost of inputs required to produce this grain has increased in Jiangsu as well as throughout China (Lardy, 1984). If costs of production for particular types of households are increasing at a disproportionate rate, then foodgrain production by such households represents a misuse of scarce capital and resources which the province cannot afford despite the high unit yields.

Findings and Analysis

The Principal Components Analysis was conducted using seven variables by which variation among the households in the sample could be identified. These seven variables were: number of persons living in the household, the area of agricultural land under contract, the gross value of off-farm sales, annual agricultural net income, annual net income from sidelines (cottage industries and husbandry), annual income from industrial employment, and income from non-industrial off-farm employment.

The PCA resulted in the identification of three eigenvectors (Table 5.1). The associated eigenvalues can be used to determine the amount of total variance explained by each vector. The three eigenvectors (components) accounted for 72 per cent of the variance within the seven variables across the entire sample. The first component had high correlations with agricultural net income (.91), off-farm sales (.86), farmland under contract (.86), and family size (.54). These loadings suggest that scores on the first component are associated with households which could be viewed as "classic" agricultural peasant households. This component was labeled "the traditional agriculture component". This first component accounted for 39 per cent of the variance within the data (Table 5.1).

TABLE 5.1: **Principal components analysis on 167 households from Jiangsu Province**

	Component 1	Component 2	Component 3
Agriculture Net Value	.91	—	—
Agricultural Off-farm Sales	.86	—	—
Farm Size (mu)	.86	—	—
Family Size (persons)	.54	—	—
Industrial Income	—	.78	—
Other Income	—	−.69	—
Income from Sidelines	—	—	.85
Variance Explained by Each Component	2.7187	1.2242	1.0889
Percentage of Variance Explained by Each Component	38.84%	17.49%	15.56%

Total Variance Explained by Components = 5.03177
Total Percentage of Variance Explained by Components = 71.88%
Source: Field Surveys.

The second component accounted for 17 per cent of the variation within the data. While agricultural variables did not load highly, industrial income (.78) correlated highly with the second component. Large scores on the second component were associated with households with industrial income, but not the other types of off-farm employment. The second component was defined as the "industrial income component".

The Third and final component which was identified through the PCA accounted for 16 per cent variance within the sample. This factor was poorly correlated with agriculturally related variables and industrially-related variables, but did not correlate with income from sidelines (.84). Sidelines are cottage industries such as raising silkworm cocoons or pearl clams, the weaving of sweaters or blankets, the manufacture of brushes, or the production of ornamental trees and shrubs. The PCA produced three component scores for each household. The three component scores for each household are transformations of the original seven variables, only now the variables are uncorrelated.

Clustering Procedure Using Derived Component Scores

The component scores were used to create household clusters which display relatively homogeneous characteristics with respect to income sources and family size. The clustering method selected was Ward's minimum variance procedure (Table 5.2) (Veek and Pannell, 1989).

TABLE 5.2: **Membership in "household type" classification determined by Ward's minimum variance**

Cluster	Prefecture			
	Nantong n=20	Nanjing n=45	Suzhou n=34	Huaiyin n=68
Industrially Focused	7*	1	32	5
(Percentage)	(35%)**	(2.9%)	(94%)	(7%)
Intermediate Group	5	19	1	24
(Percentage)	(25%)	(42%)	(3.0%)	(35%)
Agriculturally focused	8	25	1	39
(Percentage)	(40%)	(56%)	(3%)	(57%)

* Number Households in Household Type by Location.

** Percent in Household Type by Location.

Source: Field Surveys.

While any number of clusters can be stipulated for the cluster analysis, the data reflected a three group natural break. The three clusters explained 72 per cent of the variation of the 167 households based on the seven variables used for the PCA. The clusters may also be given descriptive names based on the characteristics of the households classified in each type.

The first cluster might be called the "Agriculturally Focused" cluster. These households were characterised by the lowest per capita and total household incomes, a small reported workforce, and the greatest percentage of total income derived from agricultural production. The cluster analysis placed the largest number of households in this category. Again the number of households in the sample was 167, and 44 per cent (74) were included in this household type.

The second cluster included the "Intermediate Households" where income from agriculture and industrial positions are both important. The values for the income-related variables tended to be considerably higher than those of the first cluster. The amount of contracted land by household was more similar to the households in the "agriculturally focused" cluster rather than those of the final cluster. Family size for this intermediate group, was larger than for the other groups. This middle group represented 29 per cent (48) of the sample. In general, these households appeared to be much more involved in commercial agriculture than the traditional households with the value of off-farm sales more than double those of the more traditional "agriculturally focused" households. These households seem to be committed at least in part to increasing income through expanded agricultural production represented by greater off-farm sales and larger amounts of contracted land. These households also contrast to the final group which apparently have rejected agriculture as a significant means of raising household income. It is possible that this type of household represents a transitional form between the first and third types. An alternative interpretation is that these households could also potentially become "specialised agricultural households" (*zhuanye nonghu*).

The final cluster is composed of households in which industrial income was of primary importance. In addition to this source of income, these households had more working family members, less contracted land, and minimal agricultural sales. This cluster represented a new type of peasant household which, except for location, could easily be considered to be the households of suburban industrial workers. These households, the most dramatically different occupants of the rural areas in Jiangsu, might be called "Industrially Focused" households. The "Industrially Focused" households represented 27 per cent (45) of the sample. As can be seen in Table 5.2 there is also a distinctive if somewhat mixed regional pattern of household types.

These different household types reported significantly different levels of income, different sources of that income, and different demographic characteristics (Table 5.3). The total annual income for "Industrially Focused" households is almost double that of households relying on agriculture. Per capita income for these more affluent households was 57 per cent greater than for the "Agriculturally Focused" households. The "Intermediate" group have not only procured more industrial employment, but have expanded agricultural sales as well. With respect to the goals of provincial leaders, these households might be considered most successful.

TABLE 5.3: **A comparison of income and demographic variables for three household types determined by PCA/Cluster analysis**

	Household Type		
	"Agricultural"	"Intermediate"	"Industrial"
Per Capita Income (yuan)	801.5	890.7	1394.1
Total Household Income (yuan)	3286	5077	6053
Percentage of Income from Industry	11%	19%	72%
Family size	4.1	5.7	4.4
Reported Workforce	2.7	4.4	3.5
Contracted Land (*mu*)	5.1	6.7	3.1
Agricultural Sales (*yuan*)	735	1751	158

Source: Field Surveys.

Creation of Production Budgets by Household Type

Production Budgets were created by aggregating the 167 households of the sample by their assigned household type. Again, the cropping system used for this analysis was the most common in Jiangsu, and used by each household in the research sample. It is the wheat-rice double cropping system where winter wheat is sown in the fall, harvested in the spring and followed by a summer rice crop. The values presented in the budgets (Tables 5.4 to 5.6) are the mean values for inputs and the resulting output of wheat and rice for each household type. Three budgets have been created for each household type. A budget for the wheat crop, a budget for the rice crop, and a budget for the annual cropping system incorporating both crops. These budgets represent the costs of production and resulting output for one *mu* of land. Statistically, there was little difference between the budget for the "Agriculturally Focused" households and those in the "Intermediate" category. The households in these two classes, were, however significantly different from the "Industrially Focused" households in almost every way. For the "Industrially Focused" households, the output value GVAO was lower, capital inputs higher, labour was higher, net returns lower, and the net return per *yuan* was also lower. These differences were all statistically significant at the .05 level.

TABLE 5.4: **Wheat-Rice cropping system**

	Production Budget for Wheat by Household Types		
	Agricultural Focused	Mixed Ag/Ind.	Industrial Focused
Product			
Wheat Grain (jin)	603.0	603.0	561.9
Wheat Straw (dan*)	6.0	5.9	5.3
Value (yuan)			
Grain	126.6	126.6	117.0
Straw	16.2	15.9	14.3
Gross Value	142.8	142.5	132.3
Capital (yuan)			
Fertiliser	13.0	16.8	18.0
Seed	7.5	7.5	7.0
Tillage	4.5	4.0	2.0
Farm Chemicals	2.0	2.5	3.0
Total Cap. Inputs	27.0	30.8	30.0
Labour			
Hours/Mu	61.2	60.0	61.4
Yuan/Hour	.5	.5	1.3
Yuan/Mu	31.2	30.0	78.6
NET YUAN = (Gross – Capital + Labour)	84.6	81.7	25.7
NET YUAN = (Gross – Labour)	115.8	111.7	102.3
Return per Yuan Invested	4.3	3.6	3.4
Organic Manures Applied (jin)	3412	3380	1765

3.7 Yuan = $1.00. All figures for 1 mu (1/15 hectare) of land.

Source: Field Surveys.

TABLE 5.5: Wheat-Rice cropping system

	Production Budget for Rice by Household Types		
	Agricultural Focused	Mixed Ag/Ind.	Industrial Focused
Product			
Rice Grain (jin)	997.0	980.0	966.0
Rice Straw (dan*)	9.9	9.6	9.4
Value (yuan)			
Grain	189.4	186.2	183.5
Straw	23.8	23.0	22.5
Gross Value	213.1	209.6	206.1
Capital (yuan)			
Fertiliser	18.0	17.2	19.9
Seed	5.1	5.3	3.2
Tillage	5.0	5.0	4.7
Farm Chemicals	3.0	3.0	3.0
Irrigation	4.1	5.2	7.3
Total Cap. Inputs	35.2	35.7	38.1
Labour			
Hours/Mu	72.0	75.0	84.0
Yuan/Hour	.5	.5	1.3
Yuan/Mu	37.2	37.5	107.5
NET YUAN = (Gross – Capital + Labour)	140.6	136.4	63.5
NET YUAN = (Gross – Capital)	177.9	173.9	168.0
Return per Yuan Invested	5.1	4.9	4.4
Organic Manures (Applied (Jin)	2984	2942	1432

*1 dan = 100 jin 3.7 yuan = $1.00 All figures for 1 mu (1/15 hectare) of land.
Source: Field Surveys.

The capital used for agricultural production by the "Industrially Focused" households was not converted to output at the same rate as it was for the other types of households. In addition, the land of these households was also less productive. Two related features can explain this finding. The amount of applied organic materials for these households was much lower (Tables 5.4 to 5.6). As a consequence, more inorganic fertiliser was used, and production costs were driven up. The higher production costs, in conjunction with lower yields, resulted in the difference in returns for capital invested. The production budgets for the annual cropping system (Table 5.6) indicate that once costs were deducted, the return on capital for the "Agriculturally Focused" households was 26 per cent higher per *yuan* than the "Industrially Focused" households.

TABLE 5.6: **Wheat-rice cropping system**

	Production Budget for the Annual System by Household Types		
	Agricultural Focused	Mixed Ag/Ind.	Industrial Focused
Output (yuan)			
Value of Grain	316.0	312.8	300.5
Value of Straw	40.0	38.9	36.8
Total Value	356.0	351.7	337.3
Capital (yuan)			
Fertiliser	31.0	32.0	37.9
Seed	12.6	12.8	10.2
Tillage	9.5	9.3	6.7
Irrigation	4.1	5.2	7.3
Farm Chemicals	5.0	5.5	6.0
Annual Land Tax	6.5	6.3	13.1
Total Cap. Inputs	68.7	71.1	81.2
Labour			
Hours/mu	134.0	135.0	145.0
Yuan/Hour	.5	.5	1.3
Yuan/Mu	68.3	67.5	186.1
NET YUAN = (Gross − Capital + Labour)	219.2	212.2	74.9
NET YUAN = (Gross − Capital)	287.5	279.7	256.1
Return per Yuan Invested	4.2	3.9	3.1
Organic Manures Applied (Jin)	6396	6322	3418

3.7 Yuan = $1.00 All figures for 1 mu (1/15 hectare) of land.

Source: Field Surveys.

It is counter-intuitive that the "Industrially Focused" households report greater expenditures of labour both by crop, and for the annual system. The use of high volumes of organic manures and composts generally requires greater labour inputs. One possible explanation would be the type and quality of labour. If the family land is tended by children and older persons, while the principal workers are off-farm, then labour would be less efficient and greater time would be required for crop management.

Based on an evaluation of the production budgets, it is clear that the response rates of output with respect to input vary significantly by household type. In practical terms, these results suggest that there are varying degrees of quality for the management of the wheat-rice cropping system, which appear to relate to household types. The household types were identified using income and demographic data. The commitment to efficient agriculture, in relative terms, of the "Industrially Focused" households must be questioned. It well may be that the income from agriculture comprised such a small percentage of total income (Table 5.7) that the additional income which could be accrued through more efficient production was so slight as to be unimportant. The "Intermediate" households, on the other hand, have not only increased overall incomes through greater off-farm employment but have generated greater agricultural surplus as well.

TABLE 5.7: **A comparison of income sources for three types of households in Jiangsu Province, PRC**

yuan (% of total)

	Household Type		
	"Agricultural"	"Intermediate"	"Industrial"
Total Income	3286	5077	6053
Net Agriculture	1343 (41%)	2258 (45%)	827 (14%)
Net Sidelines	681 (21%)	1165 (23%)	631 (10%)
Industry	377 (11%)	974 (19%)	4370 (72%)
Other*	885 (27%)	680 (13%)	225 (4%)

* This includes cadre work, teachers, and other government positions.

Source: 167 Field Surveys.

Discussion

There is no way of knowing how representative this sample was with respect to all households with multiple incomes in Jiangsu. Income and reported yields for these households were considerably higher than county and provincial averages. There is little question that these households were above average in many ways (Tables 5.8, 5.9). The sample was selected by agricultural bureau personnel, and these particular households were presumably chosen because officials wished to present the most positive view possible of the "modernising" countryside. Any generalisations based on the results of this research, then, must be made with caution. While no claims of a representative sample are made, it is one of the largest samples of individual households which have been collected by foreign researchers in China in recent years.

TABLE 5.8: **A comparison of yields for wheat and rice in selected counties of Jiangsu Province, PRC**

County	Grain in Kg/ha	
	Wheat	Rice
Haimen	4,575	7,350
Nantong	4,725	7,609
Rudong	4,540	7,847
Liuhe	4,043	7,127
Jiangning	3,842	7,280
Wu	4,381	7,409
Wujiang	3,900	7,375
Huaiyin	4,987	7,508
Huaian	4,813	7,502
Jiangsu Province*	3,825	6,855
China (nationwide)	2,940	5,955

* *China Agricultural Year book 1986.*
Source: County Averages from Field Surveys.

Despite the skewed nature of the sample, attention should be paid to what may be a most undesirable trend. In light of the extensive adoption of the reforms, further commercialisation and industrialisation of rural areas must be anticipated. If the newfound affluence and diversified economy which has accompanied the development of "Kotadesasi" regions to date also results in declining production efficiency by specific types of households, a rapid re-allocation of land seems in order. If the significance of income from agricultural sales continues to decline in many households, it is possible that production costs may increase needlessly as busy households reject organic fertilisers in favour of cleaner, more "modern", and less time-consuming inorganic fertilisers.

TABLE 5.9: **A comparison of per capita incomes for selected counties in Jiangsu Province, PRC**

(Values in Yuan)

County	1987 Sample Average	1986 Official Average
Haimen	n.a.	564
Nantong	960	582
Rudong	1238	526
Liuhe	954	593
Jiangning	728	646
Wu	1335	824
Wujiang	1027	806
Huaiyin	696	302
Huaian	872	355
Jiangsu Province	—	461
Total Sample Average	938	—

Source: Field Surveys; *Jiangsu Sheng Shi Shen Qiben Qingkuang 1987*, (*Jiangsu Province, City and County Basic Economic Situation 1987*); (*China Agricultural Year book 1986* p. 283).

The results of this research seem important. As the distinctions between rural and urban places becomes muted because of the "Responsibility System" and "Leave the Land but not the Township" policies, households which derive income from multiple sectors have once again become common in Jiangsu. These households will be the future residents of the "Kotadesasi" regions, and their numbers will expand as these regions expand throughout China. It seems clear from the results of this research that for some of these mixed income households, income from agriculture is no longer of paramount concern. To some extent this creates the seemingly inevitable conflict between peasants and the State. Provincial planners wish to assure a growing urban population of steady supplies to foodgrains and other agricultural products at the lowest cost. For a growing number of farm households in Jiangsu, increased incomes and high standards of living are being sought, not through crop production, but through alternate off-farm employment. Off-farm agricultural sales for some households will decline even more as households consume more and sell less, relying on other activities for capital accumulation. The wealthier households will also consume more meat which reduces market supplies of both meat and grain.

Lower gross output, possibly resulting from casual field management, coupled with a greater reliance on capital inputs has increased the costs of production for many households. Higher production costs lead to lower net returns which in turn lead to a disinclination among farm households to grow grain. The resulting shortages can lead to greater imports, the loss of foreign capital, and record levels of subsidisation in the agricultural sector (Ignatius, 1989).

The mixed income household may not be the "terminal" type for these rapidly developing "Kotadesasi" areas. These households may reflect a transitional stage in which full specialisation of labour has yet to occur. The "Specialised Household" programme which furnishes larger amounts of land, machinery, and capital to skilled farmers, who agree to grow only foodgrains and "key" economic crops, might be seen as a fundamental programme for the continued development of these areas (Wu, 1988). As the incomes of households rise with off-farm opportunities, family members may be willing to void land contracts ahead of schedule and transfer this land to full-time farmers. This would improve the scale of crop production, and transfer scarce land and agricultural inputs from the control of the less efficient "Industrially Focused" households, presumably to more willing and experienced hands.

Expansion of the "Specialised Household" programme is dependent on the availability of arable land. The land must come from households who agree to forego all but their household plots. If families are to give up their rights to land, they must be convinced that their off-farm employment is secure and reliable. Given the relationship of peasants to the land, this will be difficult. Land is consistent, even during times of tumultuous change like the present.

The growing disparity of incomes among households, even at the local level as reflected in the data, is also cause for concern. An awareness of this problem prompted the "Industry Supporting Agriculture" programme (*Yi Gong Jian Nong*) (Yu, 1987) as well as the familiar sounding slogan' "Let some prosper first so that others may follow" (*Rang yishie ren xian fugilai, ranhou zairang suoyoude ren fugilai*). "Industry Supporting Agriculture" is an important adjustment to earlier policies as a percentage of the profits from township enterprises must now be redistributed to farmers in the township who grow grain and important economic crops. In 1985, this transfer of monies included 80 million *yuan* distributed by industrial operations to full-time farmers in rural townships of Jiangsu. The purpose of this policy was to improve the parity between on-farm and off-farm incomes, and to promote grain production.

The rural reforms introduced in Jiangsu in 1979-80 have transformed many areas from primarily agricultural economies to mixed economies with greater percentages of employment in industry, service, transportation, and construction. This shift has resulted in households which derive income from multiple sources. Households with industrial positions tend to have higher incomes and, at least for the households surveyed for this research, are less efficient producers of foodgrains. The higher incomes have led to programmes designed to restore local party. The differences in production efficiency may, in part, be the impetus behind recent efforts to promote the "Specialised Household" programme. It appears that Chinese planners feel that sufficient time has elapsed to evaluate the 1979 reforms, and have already taken steps to develop appropriate adjustment policies. There is no doubt that the reforms have resulted in new types of places, neither urban nor rural, and that the employment opportunities in these areas are different as well. What is less certain is how both efficiency and total agricultural production will be affected, and who will be farming in the "Kotadesasi" of the future.

REFERENCES

Caukins, Peter, 1983 : "Agricultural Economics — U.S. and China: Is There a Meeting Ground", *Agricultural and Rural Development in China Today*, (New York: Cornell University Mimeographs), No. 102, pp. 120-139.

Chenery, Hollis and Moshe Syrquin, 1975 : *Patterns of Development 1950-1970* (London: Oxford University Press).

Cheng, Lu, 1987 : "Lun Woguo Jiangshi Shengchang Jiang Yiba Fazhan Wenti", A Discussion of the Problems of Further Grain Production in China", *Jingji Dili (Economic Geography)*, Vol. 7, No. 2, pp. 83-87.

China Agricultural Yearbook, 1987 : (Beijing Agricultural Publishers), p. 223, English edition.

Crook, Frederick W., 1986 : "The Reform of the Commune System and the Rise of the Township-Collective-Household System", in Joint Economic Committee, *China's Economy Looks Toward the Year 2000*, (Washington, D.C.: U.S.G.P.O.), pp. 354-375.

Domes, Jurgen, 1982 : "New Policies in the Communes: Notes on Rural Societal Structures in China, 1976-1981", *Journal of Asian Studies*, Vol. XLI, No. 2 (February), pp. 253-267.

Editorial Board, 1986 : *Jiangsu Jingji Nianjian* (Jiangsu Economic Yearbook) (Nanjing Jiangsu People's Publishers) pp. III-65—III-84; III-110.

Editorial Board, *Ibid* p. III-17.

Fei, Hsiao Tung, 1983 : *Chinese Village Close-up*, (Beijing: New World Press), pp. 163-177.

Fei, J.C. and G. Ranis, 1964 : *Development of the Labour Surplus Economy: Theory and Policy* (Homewood, Illinois, Irwin Publishing Co.).

Foreign Broadcast Information Service, 1982 : (Trans.) *Jiangsu Agricultural Geography*, China Report: Agriculture no. 197, (Washington: National Technical Information Service), pp. 1-59.

Ignatius, Adi, 1989 : "Bittar Harvest", *The Wall Street Journal* Vol. CCXI, No. 12, January 19, p. 1.

Jackson, Barbara B., 1983 : *Multivariate Data Analysis: An Introduction* (Homewood, Illinois: Irwin Publishing Co.), pp. 111-130.

Lardy, Nicholas R., 1983 : *Agriculture in China's Modern Economic Development* (Cambridge: Cambridge University Press), pp. 111-128.

McGee, T.G. 1989 : "Urbanisasi or Kotadesasi: Evolving Patterns of Urbanisation in Asia", in Frank Costa *et al.* (editors) *Urbanisation in Asia*, Honolulu: University of Hawaii Press, pp. 93-108.

Oi, Jean C., 1986 : "Peasant Grain Marketing and State Procurement: China's Grain Contracting System", *The China Quarterly*, No. 106, pp. 272-290.

Ray, Alice Allen (ed.), 1982 : *SAS Users Guide: Statistics* (Cary, North Carolina: SAS Institute), pp. 309-346.

Ray, Alice Allen, 1982 : *Ibid.*, pp. 417-432.

Rothenberg, Jerome, 1987 : "Space, Interregional Economic Relations, and Structural Reform in China", *International Regional Science Review*, Vol. II, No. 1, pp, 5-22.

Song, Tingming, 1986 : "Review of Eight Years of Reform", *Beijing Review*, Vol. 29, No. 51 (December 21), pp. 14-16.

Travers, Lee, 1984 : "Post-1978 Rural Economic Policy and Peasant Income in China", *The China Quarterly*, No. 98, pp. 241-259.

Veeck, Gregory and Clifton W. Pannell, 1989 : "Rural Economic Restructuring and Farm Household Income in Jiangsu, People's Republic of China", *Annals of the Association of American Geographers*, Vol. 79, No. 2, pp. 275-292. These data are described, mapped and analysed in that article.

Weins, Thomas, 1982 : "The Limits of Agricultural Intensification: The Suzhou Experience", *China Under the Four Modernisations*, (Part 1) (Washington, D.C.: USGPO), pp. 462-474.

Wu, Jinliang 1988 : "Wuxi Xian Liangshi Shengchan di Shidu Guimo: Fazhan Buzhou Ji Duice", (The Appropriate Scale for Grain Production in Wuxi County: Developing Correct Steps), paper presented at a Research Meeting, Nanjing Agricultural University (October 20).

Yu, Zhan, 1987 : "To Aid Farming with Township Enterprise Profits", *China Agricultural Year Book* (Beijing: Agricultural Publishers), pp. 42-43.

Zar, Jerrold H., 1984 : *Biostatistical Analysis*, (Englewood Cliffs, New Jersey: Prentice Hall).

Zweig, David, 1987 : "From Village to City: Reforming Urban Rural Relations in China", *International Regional Science Review*, Vol. II, No. 1, pp. 43-58.

6

PATTERNS OF AGRICULTURAL GROWTH IN BIHAR

RAMESHWAR THAKUR

AFTER INDEPENDENCE, particularly during the last two decades, there has been considerable change in almost all the parameters of agriculture in India. In order to meet the growing demand of food to feed the teeming millions, agriculture has been intensified on the one hand and, on the other, attempts have been made to bring more and more areas under cultivation, and as such there has been vertical and horizontal expansion of agriculture. From time to time the government has introduced new programmes and policies for the development of agriculture.

As a result of the new programmes and policies all the parameters of agriculture have undergone a change. The net area sown has increased considerably and the gross cropped area has almost doubled. The irrigated area has increased from about 226 million hectares (1950-51) to about 470 million hectares (1976-77), which amounts to 108 per cent increase over the base year of 1950-51. Wastelands have been reclaimed and the areas under forest and other cultivable waste have declined. The cropping patterns have changed. More and more emphasis is being given to the cultivation of high-yielding and more remunerative varieties of crops. Coarse grains are being replaced by fine grains. Thus, by and large, the monetary returns per hectare have also increased to a considerable extent.

But due to variation in physical and socio-economic conditions, these changes in agriculture are not uniform all over the country either spatially or temporally. There are States like Maharashtra, Meghalaya and Assam where very little change has taken place; on the other hand, in Haryana, Punjab, Rajasthan and Uttar Pradesh vast changes have taken place in agriculture. Similarly, during certain periods of time the change has been slow and steady, whereas during some others, a rapid change has been witnessed. This provides a very important and potential field of research in agricultural geography.

The objectives of agricultural geography are to describe and explain the variable geographical patterns of agricultural development and to discover the causes and consequences of areal distributions (Tarrant, 1974). Several eminent scholars have explained the spatio-temporal variations in agricultural growth and it will be worthwhile to briefly review them here. Prior to the 1950s, however, there was hardly any separate study on the subject. Hence such analytical studies are neither numerous nor very old.

In India and abroad many scholars of various disciplines have emphasised the need for measuring and explaining regional variations in agricultural growth and have adopted different approaches. One approach is based on the assumption that the growth in output in a region during a period of time can be measured by the change in area cultivated, and the change in yield per unit area. According to Alan Huston, the growth in acreage has made a major contribution to the total growth of crop output in an Indian agriculture during 1952-53 to 1961-62 (Huston, 1968), although it has varied widely in different States. He finally concludes that future horizontal growth would be difficult due to lack of additional culturable land which could be brought under cultivation.

Minhas and Vidyanathan (1965) have explained the inter-and intra-State variations in output in terms of variations in the extent of variable culturable area and per-hectare yield. The Government of India has also conducted several studies on inter- and intra-regional variations and inter- and intra-State variations in crop output on the basis of the model presented by Minhas and Vidyanathan. The broad conclusions of these studies correspond with those of Minhas and Vidyanathan.

According to S. K. Rao (1971), the major cause of inter-State variations in crop output is the difference in the growth of irrigated area in different States. His hypothesis has also been supported by Shangari and Gupta (1978) with the addition of the specific role of supplementary input fertilisers, high-yielding varieties and mechanisation which themselves required adequate irrigation facilities.

The role of socio-institutional and intra-structural factors in explaining regional variation in agricultural growth has been highlighted in several papers. V. Nath (1970) points out the importance of land reforms development of cooperatives and expansion of infrastructure, viz. communication and rural electrification in the growth of agriculture. He developed an index of economic development and made his analysis in the light of a general comparison of ranks.

S. Bhalla (1977) suggests the increase in agricultural output is due to improvement in infrastructure, acreage structure of landholdings and institutional factors. He concludes that variation in output growth is due to lack of various inputs and technology.

The work of Utpal Baruah (1979) is an attempt to find out the relative importance of modern technology in the development of agricultural growth in Punjab and Haryana. He uses a stepwise regression procedure to identify the influential determinants for the inter-district variations in the area under high-yielding varieties. He concludes that high-yielding varieties are very much influenced by irrigation, agricultural labour, productivity in monetary terms, as well as educated and literate population, and are least affected by soil rating index in Punjab and Haryana. The study emphasises that the role of technological factors is more than of environmental factors in determining the agricultural growth.

From the brief overview of literature, two things emerge: firstly, in most of the studies dealing with the agricultural growth, future trends, spatial patterns and the determinants of the components over a time period are not stressed. A need remains of understanding temporal changes and determinants of the components involved in the spatial organisation of the agricultural pattern. Such studies will lead ultimately to the formulation of a much-needed agricultural growth policy; secondly, very little work has been done on the agricultural geography of Bihar, particularly with reference to intra-State variations in its growth. In this research paper an attempt has been made to study the spatio-temporal variations in agriculture in the State of Bihar, keeping in view the hypotheses, assumptions and findings discussed above.

Although in Bihar, agriculture is a predominant sector of economy involving about 80 per cent of population, it has been languishing for a long period of time. Despite several attempts through various plans the process of agricultural growth has not yet been properly channelised. Thus there is a need for evolving a strategy for agricultural development incorporating guidelines for the selection of schemes, technical choice, timings and sequences of investment. Further, Bihar being a food deficit State with recurring floods, droughts and famines needs special attention to its agriculture to meet the growing demands of food.

The basic objective of this work is to describe, analyse and interpret the patterns of agricultural growth in the State of Bihar. The study further attempts to explain the regional imbalances in agriculture and also examines the spatio-temporal variations of the distribution of crop during 1963-64 to 1971-72.

Data Source and Methodology

It is true that the validity and reliability of primary data is higher than that of secondary data but since it is almost impossible to collect primary data for all kinds of landuse for the present paper for two points of time, different government and quasi-government publications* have been tapped for the secondary data used in the present study. Care has been taken to ensure the accuracy, reliability and homogeneity of the statistical information given.

In the present study, district has been selected as the unit of study. Although, there are, at present, 31 districts in Bihar the existing number of districts in 1963 was 17. So only there 17 districts have been taken into account in order to avoid the problem of splitting data. An adjustment has thus been necessary for the 1972 data. Four districts, viz. Hazaribagh/Dhanbad and Monghyr/Saharsa have been combined into two owing to boundary changes and mutual transfer of area from one district to another. For a comparative study of this kind it was not possible to make a study of the new districts that have come into existence, since no data is yet available for them.

Two points of time have been selected: 1963-64 and 1971-72. To minimize any anomalies arising out of a fluctuation in the area and output of crops an average of three years (1962-63, 1963-64, 1964-65) for 1963-64 and (1970-71, 1971-72, 1972-73) for 1971-72 was worked out for each point of time. The choice of these years was made because 1962-65 are the pre-Green Revolution years, whereas, 1970-73 are the

* These include area and production of crops under study from the *Annual Season and Crop Reports*, Bihar; *Agricultural Situations in India; Estimate of Area and Production of Principal Crops in India; Indian Agricultural Statistics; Bihar Statistical Handbooks and Fertiliser Statistics.*

post-Green Revolution years when a new technology in Indian agriculture made a large impact.

The following 17 crops both foodgrain and cash were selected: rice, wheat, maize, jowar, bajra, ragi, barley, gram, tur, rape and mustard, sesamum, linseed, castorseed, cotton, jute, sugarcane and tobacco. These crops account for about 80 per cent of the total cropped area in the State, the rest 20 per cent constituting miscellaneous crops has not been taken into account.

Two parameters of agriculture, viz. area and production, have been considered. In order to study the spatio-temporal variations of different crops, the annual compound growth rate in per cent has been calculated. The districtwise overall change in area and production was also calculated. Production value has been converted into monetary value which may be used to judge the agricultural production on an equal footing. The average price of all the crops at State level was chosen for the base year 1963-64 and has been applied for converting the production of various commodities in monetary terms. The same price rate has also been used for converting the agricultural product of 1971-72 in order to maintain uniformity in the price level. The obtained monetary value of all the crops has been arranged in descending order and a suitable grade of growth has been marked. Finally, a functional relationship between the growth of area, production in terms of monetary value and associated variables has been established by means of correlation-regression analysis. The dependent variables are the growth rates of area and monetary value of agricultural product, and the independent variables are irrigation, fertilisers, mechanisation tractors and pumping sets. Pearsonian product moment correlation coefficient has been used for finding out the strength of relationship between them.

The Study Area

The State of Bihar is a compact quadrilateral territorial unit in the north-eastern part of India. It lies between Latitudes 22°N and 27° 31′N and Longitudes 83° 20′E and 88° 17′E and is bounded by Nepal in the north, Orissa in the south, Uttar Pradesh and Madhya Pradesh in the west and West Bengal in the east. It covers an area of 174,038 sq.km. which constitutes 5.71 per cent of the total area of India. Among the Indian States, Bihar occupies eighth place in area and second in population. The total population of Bihar is 56,353,369 which gives a density of 324 persons per sq. km. (1971). This is much more than the average density of population in India (179 persons per sq. km.)

Bihar has distinct characteristics of structure, physiography, climate and soil, which combined together give it a sense of individuality. Physically, the State can be divided into three parts : (a) Himalayan Foothills, (b) Ganga Plain, and (c) Chotanagpur Plateau. The Ganga Plain is the most important agricultural tract of the State owing to its homogeneous relief, an altitude of less than 500 ft. above sea level, with a gentle slope, fertile alluvial soil and favourable climatic condition, whereas, the Chotanagpur Plateau and the Himalayan Foothills region rising to over 2,000 ft. above sea level are less significant from the agricultural viewpoint. The drainage system is dominated by the Ganga along with its tributaries and distributaries in the plains, while Damodar and Subernarekha are the predominant rivers in the Chotanagpur Plateau. Due to seasonal variability of rainfall, together with relief the plateau rivers are largely precarious and turbulent in nature. Frequent floods in the plains cause tremendous damage to the standing Kharif crops and sometimes delays the sowing of Rabi crops. Climatically, the State enjoys a hot and humid monsoonal climate. The distribution of annual rainfall varies from 191 cm. near the north-eastern corner to a minimum of 102 cm. near the western border of the State. There is a considerable variation in the range of temperature: annual, seasonal, monthly and diurnal. The rainfall is highly seasonal in concentration and variable in character, which affects the agriculture to a considerable extent. There are a large variety of soils found in the State but the major ones are : (a) swampy soil, (b) alluvial soil, (c) calcarious soil, (d) forest soil, (e) red soil, (f) mixed red and black soil, and (g) lateritic soil. Red soil is mostly found in the Chotanagpur Plateau while alluvial is extensively found in the Plain areas.

Bihar is essentially an agricultural State where more than nine-tenths of the total population depends on agriculture, directly or otherwise. The landuse pattern is influenced by the complex physical and socio-economic factors. Sixty four per cent of the total area is cultivable and the remaining 36 per cent is given to non-agricultural uses including forest, groves and pastures. The percentage of cultivated land (net area sown) varies considerably in different districts, depending on the degree of flat land available. It ranges from 52 to 73 per cent in the South Bihar plain, 67 to 72 per cent in the North Bihar plain and less than 40 per cent in the Chotanagpur Plateau. The proportion of forest cover also varies greatly. It is negligible in the plain districts but covers a large portion of the south-western corner of the plateau region. The percentage under fallow land also varies. It is comparatively higher in the Chotanagpur region (above 20%) than in the Bihar plain (below 12%). Land under culturable waste is similarly high all over the Chotanagpur region and low in the Bihar plain area. Land not available for cultivation is generally occupied by settlement roads, railways, water bodies and barren lands. It ranges from 10 to 23 per cent throughout the State.

Measurement of Agricultural Growth

For the measurement of agricultural growth, a number of parameters of agriculture such as area, yield production (actual and its monetary value), input-output ratio, levels of mechanisation and many others could be taken. But in the present study, owing to lack of time, resources and paucity of reliable secondary data only two parameters — growth in area in various components of agriculture and growth in total output — have been considered. In order to make the production of various commodities comparable, it has been converted into monetary value. The growth of agriculture in the area has been studied at four levels so that a clear and composite picture may emerge : firstly, the changing patterns of general landuse with special reference to fallow lands and net area sown; secondly, changing patterns of crop landuse during both the agricultural seasons, viz. Kharif and Rabi; thirdly, cropwise growth in area and output; and fourthly, overall growth at the State level.

Changing Patterns of Landuse

As a result of various geographical factors there is considerable variation in landuse patterns both in time and space. The proportion of forest cover is found to vary greatly. Most of the plain districts have a negligible forest cover, while large parts of the Chotanagpur plateau are densely forested. This may be ascribed to rugged relief coupled with moderate rainfall in the plateau region, and a flat, fertile and homogeneous land in the plain region. These geographical factors have made a distinct impact on the landuse change and proportion in the State.

The average area under various landuse categories in Bihar and their percentage to total area during the agricultural years of 1963-64 and 1971-72 give an interesting pattern. The total area of the State is 17,330,269 hectares. In 1963-64, the area under forest, land put to non-agricultural uses, fallow lands, other uncultivated and excluding fallow lands and net area sown was 3,708,335 hectares (21.4%), 2,276,711 hectares (13.2%), 1,900,298 hectares (11.0%) 8,98,222 hectares (5.1%), 8,546,703 hectares (49.3%), respectively which has correspondingly become 2,796,703 hectares (16.1%), 2,624,131 hectares (15.1%), 2,749,553 hectares (15.9%), 8,83,400 hectares (5.1%), 8,276,462 hectares (47.8%) in 1971-72. It is clear that an increasing trend has been found in the categories of land not available for cultivation, gross cultivated land and fallow land, while a declining trend has been recorded in the proportion of forest and net area sown. The districts located in the plain and plateau areas have displayed some contrasting changes. A declining trend is observed in other uncultivable land excluding fallow land and net area sown in the plain districts, while an increasing trends has been found in both the categories in the plateau districts (Table 6.1). Thus, plateau districts are gaining more land under net area sown at the cost of declining forest area, while plain districts are losing land proportions under net area sown due to waterlogging, extension of saline and alkaline lands or practices of leaving the land fallow due to lack of irrigation in the uplands and fear of floods in the lowlands.

Changing Patterns of Crop Landuse

The agriculture year is divided into two main seasons — Kharif, the season of summer crops, and Rabi, the season of winter crops. The Kharif season includes the *bhadai* and *aghani* crops starting with the advent of the monsoon, generally in the second week of June. The principal Kharif crops are rice, maize, ragi, jowar, bajra, tur, sugarcane, sesamum, cotton and jute. These crops are generally produced without irrigation, with the exception of sugarcane. Rabi is known as the season of winter crops which are sown in October and harvested in March and April. The crops of this season, are wheat, barely, gram, linseed, rapeseed and mustard, castorseed and tobacco. As during this season the rainfall is meagre and highly variable, most of the Rabi crops rely either on the moisture available in the soil after the monsoon rains or irrigation from various sources. It is no doubt

TABLE 6.1: Changing patterns of landuse in Bihar (districtwise)

District	Forest		Land not available for cultivation		Fallow land		Other cultivable land excluding fallow land		Net area sown	
	1963-64	1971-72	1963-64	1971-72	1963-64	1971-72	1963-64	1971-72	1963-64	1971-72
Bhagalpur	11.1	8.4	15.3	22.0	11.0	12.7	3.6	4.3	59.0	52.6
Champaran	7.5	9.7	12.3	14.3	5.7	6.0	5.2	2.4	69.3	67.6
Darbhanga	—	—	15.8	18.0	10.2	9.1	4.0	3.0	70.0	68.9
Gaya	18.0	12.3	15.7	14.8	6.0	12.3	3.1	2.4	57.2	59.2
Muzaffarpur	—	—	10.7	15.2	6.8	8.0	3.6	3.2	78.9	71.6
Palamau	50.0	44.7	7.2	10.1	10.4	19.7	4.6	3.3	28.0	22.2
Patna	0.8	0.8	28.7	17.9	2.3	6.4	0.6	0.9	77.6	73.0
Purnea	1.4	0.2	17.2	15.5	13.1	17.5	4.6	3.8	63.7	63.0
Ranchi	26.9	22.1	10.3	10.9	20.0	20.7	4.0	5.1	38.8	40.9
Santhal Parganas	23.7	9.7	12.3	14.3	16.4	25.5	9.8	11.7	37.8	38.8
Saran	—	—	10.7	15.0	10.6	14.4	2.7	3.6	76.0	67.0
Shahabad	22.1	16.3	11.1	12.8	2.6	4.1	1.0	0.9	63.2	65.9
Singhbhum	46.8	24.6	11.4	16.5	13.6	20.2	3.7	7.9	24.5	30.5
Dhanbad + Hazaribagh	31.8	24.0	19.5	23.3	14.1	20.2	7.5	9.0	27.1	23.4
Monghyr + Saharsa	4.6	6.9	19.7	17.3	14.1	18.6	9.0	5.0	52.6	52.1
Bihar	21.4	16.1	13.2	15.1	11.0	15.9	5.1	5.1	49.3	47.8

Source : *Annual Season and Crop Reports of Bihar*, Patna, 1963-64 and 1971-72, pp. 3-9 and 4-12.
Note: Figures indicate percentages to the total area of each district.

difficult to increase the net area sown, but to intensify cultivation the same field can be cropped in both the seasons with suitable inputs. However, during the study period, there were some changes in the total cropped acreage strength. The area under Kharif crops had decreased from 8,545,976 hectares (79.6%) in 1963-64 to 8,470,599 hectares (76.8%) in 1971-72. But the area under Rabi crops had slightly increased from 1,714,811 hectares (20.4%) to 1,962,705 hectares (23.2%). Interestingly, the percentage increase in the cropped area during the Rabi season is exactly equal to the percentage decrease in the cropped area during the Kharif season.

A brief glance of the agricultural pattern of Kharif and Rabi seasons will indicate that during the study period the area under cereals increased, whereas the area under non-cereals decreased considerably. It would be worthwhile to discuss the growth in area, output and yield of major crops in detail.

Area, Output and Major Crops

Rice

Rice is the most important grain crop in the State in both acreage and production. It occupies nearly 47 per cent of the total cropped area and is harvested in three seasons—summer, autumn and winter. It is cultivated in every part of the State. Nearly two-thirds of its area lies in the Ganga Plain and only about one-third is in the Chotanagpur Plateau. Deep fertile clayey or loamy soils coupled with an average annual rainfall of 127 cm. are considered ideal for rice cultivation. During the study period the area under rice decreased from 47.69 per cent to 46.94 per cent and the decreasing tendency was noticed more in the Plateau than in the plain districts. Similarly production also decreased accounting to –0.73 per cent annual compound growth rate.

Maize

Maize is another important food crop grown as a rainfed Kharif crop in Bihar. It is grown on the upland which is immune to early floods. Maize is mostly concentrated in the south and west of the Burhi Gandak river covering the districts of Saran, South Champaran, South Muzaffarpur, South Darbhanga and North Monghyr. Maize crop has experienced an increase in areas in all the districts except in Purnea,

Santhal Parganas and Singhbhum, but has showed a decline in production in all the district except in Champaran, Gaya, Muzaffarpur and Patna. The notable causes for this decline are probably caused by the poor quality of seeds, low facility of irrigation; lack of mechanisation and, other technological factors.

Ragi

Among millets, ragi is the most important crop in Bihar and is sown when the rain starts. Its cultivation is largely concentrated in the districts of Darbhanga, Hazaribagh, Saharsa, Muzaffarpur, Ranchi, Purnea, Gaya and Saran. The Kosi-affected districts of Darbhanga and Saharsa have a considerable acreage under this crop. This crop is commercially not very important, and except Purnea all other districts have recorded a decline in both area and production.

Jowar and Bajra

The production and yield of jowar and bajra are the lowest among the foodgrains in the State, yet a majority of districts both from plateau and plain areas have experienced an increase in its area. The annual compound growth of production has not given any pattern which could be generalised.

Tur

Tur is a favourite crop among the pulses grown. It is grown on light soil even without irrigation. Its cultivation is done in almost all the districts of Bihar but the greater part is shared by Saran, Palamau, Muzaffarpur, Gaya and Darbhanga. It is also extensively grown in the districts of Chotanagpur. The area under tur has also declined in all the district except in Champaran and Patna and production has also declined in all the districts except in Champaran, Gaya and Patna.

Sugarcane

Sugarcane is one of the important cash crops in Bihar. It is extensively grown in the areas having well-drained calcareous soil and flood-free uplands comprising Champaran, Saran, Muzaffarpur and Darbhanga districts of the North Ganga plain and central irrigated tracts of

Shahabad, Gaya and Patna districts of the South Ganga plain. Though sugarcane has been a major cash crop still its area and production have decreased in all the district except in Champaran.

Sesamum

Sesamum is a kind of oilseed which is used for cooking purposes in many parts of the State. It requires moderate rainfall and a temperature of not more than 21°C, and is grown in well-drained loamy soil. Palamau district is most favourable for its production where the area and yield per hectare are very high as compared to other districts. The area and production of sesamum also declined during the study period.

Cotton

Cotton cultivation is not very popular in Bihar. Yet its area and production are on the increase. Patna, Shahabad, Saran, Darbhanga and Gaya are the important districts producing cotton. The area, production and yields of cotton increased considerably in Bihar during the study period.

Jute

Jute is also an important cash crop producing a cheap fibre used as raw material in the jute industry. It is cultivated in north-eastern Bihar comprising Purnea, Saharsa and Darbhanga districts. In the majority of the districts the area as well production have declined.

Wheat

During the study period wheat emerged as the second most important foodgrain crop in the State. During the early sixties there were only two districts growing wheat and they shared more than 45 per cent of the land under this crop. By the seventies, six districts recorded more than 10 per cent of annual compound growth in area and 20 per cent in production. It is notable that the output of wheat has increased in all the districts in the area under study.

Barley

Barley is another important Rabi crop grown in Bihar. Because of the extension of wheat, both the area and production of barley have considerably declined in all the districts except in Santhal Parganas.

Gram

Gram is the most important crop among pulses in Bihar. Its cultivation is mostly concentrated in the South Bihar plain comprising Patna, Gaya and Shahabad districts. The overall area and production of gram have declined in all the districts except in Ranchi and Singhbhum.

Rapeseed and Mustard

Rapeseed and mustard are the important oilseeds cultivated in winter. They provide edible oil and oilcakes. Though they are equitably produced in every district of Bihar the area has declined in Gaya, Muzaffarpur, Purnea, Ranchi, Santhal Parganas, Singhbhum, Monghyr and Saharsa. Surprisingly, however, the total output has increased in the State.

Linseed and Castorseed

These are equally important among the oilseeds grown in Bihar. The area under linseed and castorseeds has declined but the production has increased during the study period.

Tobacco

Tobacco is another important cash crop in the State after jute. Though it fetches a good price, the area as well as production declined during the study period in the State. This is because it has to compete with crops like jute which give a higher return. Moreover, there is much fluctuation in the tobacco price which affects its cultivation adversely.

Overall Growth in Area and Production (Districtwise)

The detailed analysis of the positive and negative growth in area and production of individual major crops during the period under study at district levels fails to give a clear pattern of change in agriculture for the State as a whole. Hence, the growth in area and production of all the crops considered have been added and an overall growth of agriculture in each district has been worked out. The resulting figures of area and production have been arranged in descending order and have been graded into high, low, negative and extremely negative as given in Table 6.2 (area) and Table 6.3 (production) and mapped in Fig. 6.1 and Fig. 6.2 respectively.

TABLE 6.2: Levels of agricultural growth in area

Percentage compound growth per annum	Total no. of districts	Name of the District
High > 1	2	Champaran and Singhbhum
Low 0 to 1	6	Bhagalpur, Darbhanga, Muzaffarpur, Purnea, Ranchi and Shahabad
Negative –1.5 to 0	4	Palamau, Patna, Saran and Monghyr + Shaharsa
Extremely Negative < –1.50	3	Gaya, Santhal Parganas and Dhanbad + Hazaribagh

Source : Calculated by the Author.

Growth of Area

Table 6.2 clearly indicates that out of 15 districts eight have recorded an increase while seven show a decline in area. However, it would be worthwhile to discuss each growth region separately in detail.

High Growth Region

The districts having more than one per cent of annual compound growth have been kept under high grade. Champaran lying in the extreme north-western corner and Singhbhum located in the extreme south are the only two districts which have shown a high increase in the cropped area in the State (Fig. 6.1). This comparatively higher growth is mainly due to two reasons, firstly, clearing of forest cover to bring it under cultivation; and secondly, reclamation of some of the uncultivated land for increasing the cultivation of crops.

Low Growth Region

In the districts falling under this category, the percentage of growth in area is insignificant and varies from zero to one per cent. There are two contiguous regions of low growth : one is located in the north-eastern

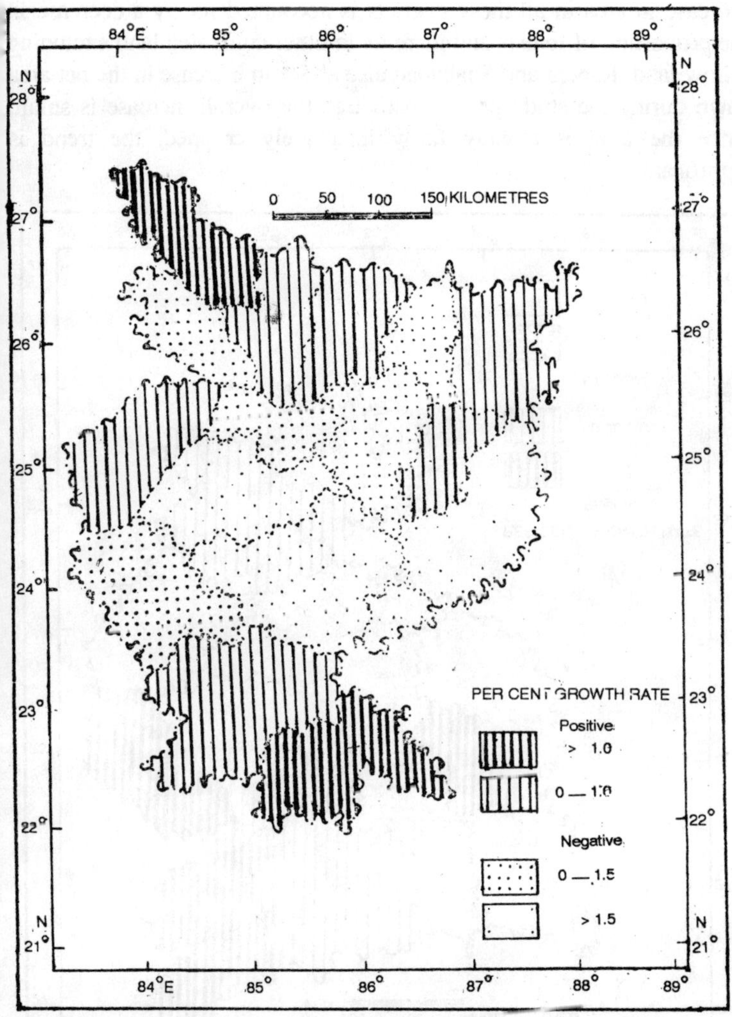

FIG. 6.1: **Bihar—levels of agricultural growth—1963-64 to 1971-72 (area).**

part interrupted by Saharsa district, while the other lies in the south-western part separated by Palamau district. The former includes the districts of Muzaffarpur, Darbhanga, Purnea and Bhagalpur where the growth in area is 0.32, 0.54, 0.13 and 0.26 per cent respectively, while the latter includes the districts of Ranchi and Shahabad where the growth rate is only 0.21 and 0.08 per cent respectively. The increase in area in all these districts is accounted for by a decrease in the proportion of forests and increase in other cultivated land excluding fallow land. Ranchi and Shahabad also show an increase in the net area sown during the study period. Although the overall increase is small, since the land is already fairly intensively cropped, the trend is significant.

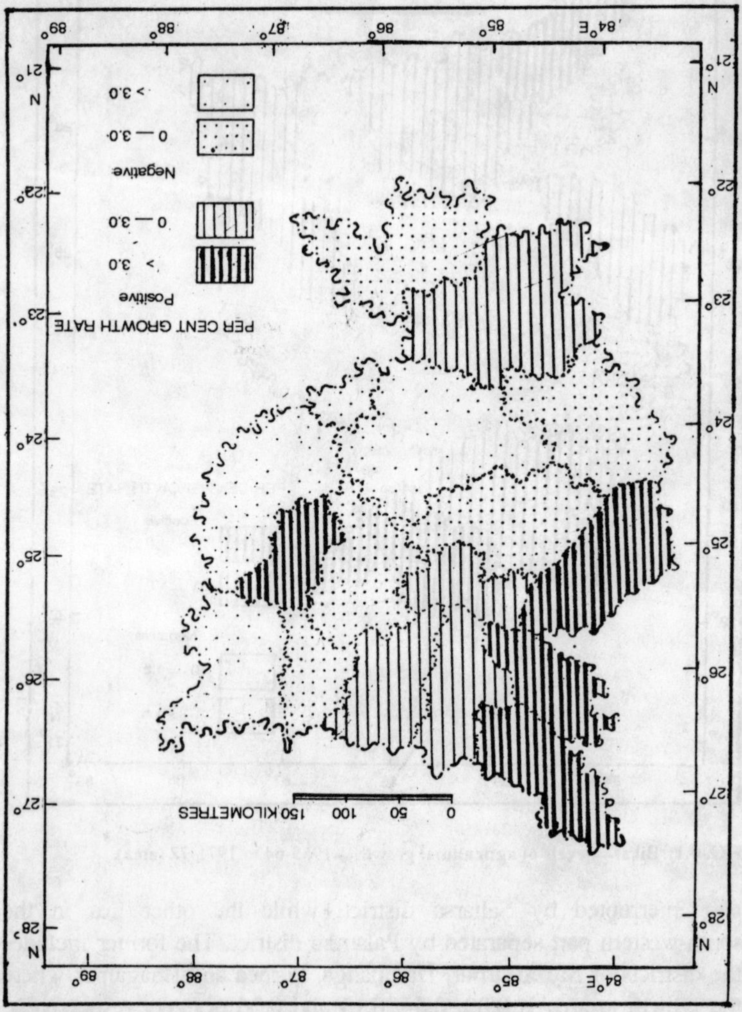

FIG. 6.2: Bihar—level of agricultural growth 1963-64 to 1971-72 (production).

Negative Growth Region

Figure 6.1 showing the levels of areal growth of agriculture portrays a very interesting spatial pattern. With the exception of Saharsa, the region of negative growth lies in the central belt of the State running from east to west and the region of positive growth occupies the northern and southern belts. However, out of 15 districts there are seven where the area under cultivation has declined. These districts are marked by comparatively less facilities of irrigation, fertilisers and mechanisation. Although there are several causes for negative growth, the important among them are the decrease in net area sown, increase in fallow land and the deterioration of soil fertility. It is clear from the landuse data that the districts of plain area where cultivated area has declined are marked by a high percentage of the increase in fallow land as compared to the plateau region. This is because of the variability of rainfall and the frequent occurrence of floods. The districts which have recorded a decline are categorised into two grades—negative and extremely negative. Palamau (10.7%), Patna (0.64%), Saran (0.08%) and Monghyr and Saharsa (0.09%) are placed in the first grade while Dhanbad and Hazaribagh (2.21%), Gaya (1.57%) and Santhal Parganas (1.60%) fall into extremely negative grade. All these districts lie south of the Ganga, and are marked by a relatively lower rainfall which corresponds to a higher variability of rainfall.

Growth in Production

Table 6.3 shows the change in production in different districts. It is clearly portrayed again that out of 15 districts, eight have registered an increase, while seven show a decline in production. The percentage of growth has thus been categorised into high, low, negative and extremely negative grades and are mapped in Fig. 6.2 which shows an interesting spatial pattern. The region of positive growth forms a contiguous block with the exception of two districts, viz. Bhagalpur and Ranchi, which have recorded a low increase in area. The contiguous region lies in the north-western portion of the State. However, it will be worthwhile to discuss the growth pattern of each region in detail.

TABLE 6.3: Levels of agricultural growth in production

Percentage compound growth per annum	Total No. of Districts	Name of the District
High > 3	4	Bhagalpur, Champaran, Saran and Shahabad
Low 0 to 3	4	Darbhanga, Muzaffarpur, Ranchi and Patna
Negative —3 to 0	4	Gaya, Palamau, Singhbhum and Monghyr + Saharsa
Extremely Negative < —3	3	Purnea, Santhal Parganas and Dhanbad + Hazaribagh

Source : Calculated by the Author.

The Region of High Growth

The districts whose production has increased by more than three per cent have been kept under the category region of high growth. The maximum growth is recorded in Saran district (4.37%) followed by Shahabad (3.52%), Bhagalpur (3.25%) and Champaran (3.04%). The increase in production depends on the interaction of several institutional, technological and environmental factors. Due to lack of detailed data it has not been possible to fully analyse the factors that are responsible for high level of growth or lack of growth elsewhere but it is obvious from the emerging spatial pattern that the areas with a high rate of growth are very positively associated with high levels of growth in irrigation, fertilisers and mechanisation. Except Bhagalpur other districts which have recorded a high growth rate are located in the western portion of the State, where the incidence of irrigation is higher.

The Region of Low Growth

Except Ranchi, the region of low growth forms a contiguous belt of three districts located in the centre of the northern part of the State. Between zero to three per cent growth in production is categorised in low grade. Darbhanga, Muzaffarpur, Patna and Ranchi have recorded 0.48, 1.19, 1.32 and 1.53 per cent growth rates respectively. The notable causes for this increase are an increase in area, changing cropping pattern, increasing facility of irrigation, fertilisers and mechanisation. These districts have also recorded a very low percentage of increase in area as well as in production.

The Region of Negative Growth

The region of negative growth forms a contiguous belt living in the south and south-eastern parts of the State. Its contiguity is interrupted by the districts of Ranchi and Bhagalpur which have recorded positive growth. Table 6.3 also indicates that seven out of 15 districts have a negative growth rate in production during the study period. The range of decline is between zero and 3.76 per cent. The maximum decrease is found in Dhanbad and Hazaribagh districts where both area and production have declined by 2.21 and 3.76 per cent respectively. The other districts where there is a decline in production are Gaya (0.34%), Palamau (0.23%), Singhbhum (0.78%), Monghyr and Saharsa (0.92%). Besides these, Purnea (3.67%); Santhal Parganas (3.07%) and Dhanbad and Hazaribagh (3.76%) have been placed in extremely negative group (Fig. 6.2). Although there are several causes of negative growth, the important among them include deterioration of soil, fertility, reduction of area, changing emphasis of crop cultivation, lack of adequate irrigation facilities, fertilisers, manures and improved seeds.

Tables 6.2 and 6.3 further reveal that the production is directly related to the area under a crop. For example, out of 15 districts, six districts, namely, Bhagalpur, Champaran, Darbhanga, Muzaffarpur, Ranchi and Shahabad have displayed an increase in both area and production, while five namely, Gaya, Palamau, Santhal Parganas, Dhanbad and Hazaribagh, and Monghyr and Saharsa show a decline in both. Four districts including Patna, Purnea, Saran and Singhbhum have not shown any corresponding trend. It is, therefore, clear that if the cropped area is increased in the State, production would automatically increase.

Overall Growth in Monetary Value

The earlier analysis of the crop pattern has been done in terms of area and weight. But the production value of various categories of crops cannot be compared for gaining an insight into the changing patterns of growth. For example, the production of rice, wheat and other crops is generally measured in terms of tonnes, whereas, cotton and jute are given in terms of bales. Even if the product is measured in the same unit of weight it is not proper to compare various products having different monetary values. For example, it is meaningless to compare one kilogram of wheat or rice with one kilogram of opium or coffee. Thus, a problem arises in the comparison of the total value of all the crops. The alternative solution to this effect is to convert the production of all crops into a uniform unit so that a standard and comparable result can be achieved. This has been done in the following quantitative analysis for making a generalisation.

For converting production into a single comparable unit, monetary value has been selected. Monetary value is the only standard value which could be used to judge the agricultural production on an equal footing. Firstly, the prices of all crops at district level were obtained but they had to be rejected owing to their variability and non-comparable character. Alternatively then, the uniform price list of the crops on a State basis has been selected, so that uniformity may be maintained. The price list has been chosen for the base year only, i.e. 1963-64. It represents the average prices of 1962 to 1965. The same price list has been used for the respective crops of 1971-72. On this basis, firstly, the production of all crops of each district has been converted into monetary value for both points of time separately; secondly, the monetary value of all the crops has been added up for individual districts; and finally, the compound growth rate for each district has been calculated. The results are arranged in descending order and graded into four categories as given in Table 6.4 which clearly indicates that out of 15 districts, eight have recorded an increase, while seven show a decline in total monetary value.

TABLE 6.4: Levels of agricultural growth in monetary value

Percentage compound growth per annum	Total No. of Districts	Name of the District
High > 3	3	Bhagalpur, Saran and Shahabad
Low 0 to 3	5	Champaran, Darbhanga, Muzaffarpur Patna and Ranchi
Negative –2 to 0	5	Gaya, Palamau, Purnea, Singhbhum and Monghyr + Saharsa
Extremely < –2 Negative	2	Santhal Parganas and Dhanbad + Hazaribagh

Source : Calculated by the Author.

The Region of High Growth

The region of high growth lies in the contiguous belt in the western part of the State except the district of Bhagalpur which lies in the eastern part. The districts having recorded over three per cent annual compound growth rates in monetary value have been included under this grade. These are Saran, Bhagalpur and Shahabad which have recorded an increase of 4.53, 3.31 and 2.85 per cent compound growth rate respectively. The notable causes for this growth are mainly high yield rate and increase in production caused by better facility of irrigation, fertilisers and other technological factors.

The Region of Low Growth

The region of low growth is really a continuation of the region of high growth except in the district of Ranchi and is located in the north-western part of the State. The districts of Champaran, Darbhanga, Muzaffarpur and Ranchi are included in this region and have recorded comparatively a low growth ranging between zero and three per cent. The reasons ascribed are low increase in area and production. Except Ranchi the other districts of the Chotanagpur region have recorded a negative growth in monetary value.

The Region of Negative Growth

Table 6.4 indicates that seven districts have shown a negative growth during the study period. The range of decline is between zero and 3.77 per cent. This region exactly corresponds with the region of negative growth in the area. Dhanbad and Hazaribagh and Santhal Parganas have recorded a decline of over one per cent and rest of the districts have recorded below one per cent decline. Although there are several causes of negative growth, the important among them are a sharp decrease in area and production during this period, which is either due to the diversion of better quality land to crops of higher yield or weather abnormality, since southern Bihar is generally prone to drought and erratic nature of rainfall. The districts where the growth has declined by zero to two per cent have been placed in the negative group. These are Gaya (0.53%), Palamau (0.17%), Purnea (0.89%), Singhbhum (0.48%) and Monghyr and Saharsa (0.78%). The remaining two districts, viz. Santhal Parganas (2.39%) and Dhanbad and Hazaribagh (3.78%), have been placed in the extremely negative group.

Determinants of Agricultural Growth

The above analysis primarily covers the spatial and temporal patterns of agricultural growth and decline in the area and production of major crops grown in Bihar. It is clear that eight districts—Bhagalpur, Champaran, Darbhanga, Muzaffarpur, Purnea, Ranchi, Shahabad and Singhbhum—have recorded an increase in both area and production while three—Patna, Saran and Shahabad – have registered a growth only in production. We may, thus, attempt to analyse quantitatively the determinants of agricultural growth and establish the functional relationship between the area and production and associated variables by means of correlation-regression analysis. Specifically, this investigation attempts by means of hypotheses formulated in quantitative terms to account for spatial differences in the amount of agricultural growth which occurred in the 15 districts in Bihar during 1963-64 and 1971-72.

The Dependent Variables

In this study two dependent variables have been used. The dependent variables or Y co-ordinates for each unit are the spatial variations in the growth rates of area and production. The districts are the unit areas for which data of area and production for two time periods have been collected. Separate growth rates for each crop during 1963-64 and 1971-72 have been identified and the total increase or decrease for each district has been determined. Agricultural growth varies from district to district as can be seen from Fig. 6.1 and Fig. 6.2. These maps indicate that districtwise differences in agricultural growth exist and the recognition of these differences aids in understanding the spatial pattern of agricultural growth. This variation can be explained, and permits the testing of a large number of hypotheses.

The Independent Variables

Agricultural growth is affected by diverse factors and they differ from region to region. The unequal agricultural growth in different districts of Bihar demands investigation of the factors influencing them. From the findings of previous studies as well as from observation it is ascertained that there are three principal factors besides the physical factors, which have contributed a lot to the growth of agriculture in Bihar. These three independent variables are irrigation, fertiliser and mechanisation (tractors and pumping sets) which have been selected for an statistical analysis. It is hypothesised that each of these factors is really associated with agricultural growth and decline.

It is widely assumed that one type of economic activity, viz. irrigation, is particularly important in accounting for agricultural growth and decline. In previous studies also irrigation is known to bear the highest correlation with productivity (Khusro, 1968). Irrigation is indispensable for raising a crop in the dry season all over the State. A comparatively analysis of levels of agricultural growth and an increase in the irrigated area indicate that by and large agriculture has shown a positive growth in the districts where the irrigated area has increased. It is hypothesised that areas having a higher growth in the area of irrigated land would experience a large amount of growth under area and production than those having lesser growth in the area of irrigated land.

Fertilisers, the modern chemical input of the scientific age, is also one of the important elements determining the development of agriculture. Fertilisers are applied to increase the nutrient status of the soil and to maintain it at a high level of productivity from year to year. The spatial pattern of the consumption of fertilisers in Bihar has a great impact on the distribution of crops. A comparative study of levels of growth of agriculture and increase of consumption of fertilisers shows a close relationship between the two variables. It is, therefore, hypothesised that those areas in which application of fertilisers is higher would attain higher levels of agricultural growth than those in which it is lower.

Modern agriculture is very much affected by the use of improved machinery and implements. Although a number of improved machineries such as tractors, pumping sets, harrowers, winnowers, weeders and harvesters are used in agriculture in Bihar, only tractors and pumping sets have been considered in the present study because comparatively authentic and published data are available only for these two variables for both points of time under examination. The use of tractors as one of the improved machinery has increased considerably during the last decade but the increase has not been uniform all over the State. Thus in the present study it is hypothesised that those areas with a greater number of tractors used would experience more agricultural growth than areas with a smaller number.

The use of electric pumping sets in place of the traditional method of irrigation has been indicated in the literature as relevant in accounting for agricultural growth (Ramachandra, 1975). Under mechanisation this is a recent innovation in the practice of irrigation. After the Green Revolution in agriculture the number of pumps has increased rapidly in all the districts of Bihar mainly due to a better supply of electricity. Almost all the districts of the Bihar plain have recorded a high annual compound growth rate in the number of pumping sets. It is, therefore, hypothesised that those areas in which the number of pumping sets is larger would show a higher agricultural growth than areas in which the number of pumping sets is smaller.

The logic underlying the above hypotheses is that the variation which has been observed from district to district in the rate of agricultural growth is attributable to the co-variation of the factors mentioned in the hypotheses. In other words, some form of a functional relationship is understood to exist between the dependent variables, rate of agricultural growth under area and production and the independent variables. In this situation correlation-regression analysis can be used to test the adequacy of the hypotheses.

Analysis of Results

Table 6.5 represents statistics related to dependent and independent variables in which Y_1 and Y_2 are dependent variables, namely, growth rate in area and production (in terms of monetary value) respectively. Similarly X_1, X_2, X_3 and X_4 are independent variables, namely, irrigated area, fertilisers, mechanisation—tractors and pumping sets respectively. The irrigated area is measured in hectares, fertilisers in tonnes, tractors and pumping sets in number and all are expressed as per unit in annual compound growth. The data for independent variables are assumed to be normally distributed as well as randomly drawn from hypothetical universe. Therefore, correlation-regression techniques can be applied to determine the relationship between the dependent variable and each independent variable.

In order to correlate the growth rate in area with the four independent variables, Pearsonian product moment correlation coefficient has been calculated. The "*r*" value has been found to be 0.30 with irrigated area; 0.38 with fertilisers; 0.03 with tractors and

0.12 with pumping sets (Table 6.6). These values are very low and have been found to be not significant at .05 probability level. It would be safe to assume that they are not intimately correlated because a total of only 25 per cent of the total variation in the dependent variable has been explained (r^2). The correlation coefficient in the case of tractors and pumping sets are extremely low to have any significance while the correlation coefficient in the case of irrigated area and fertilisers seems to be weakly related.

TABLE 6.5: **Dependent and independent variables of agriculture**

District	Y_1	Y_2	X_1	X_2	X_3	X_4
Bhagalpur	0.26	3.31	1.07	24.98	11.20	42.36
Champaran	1.99	2.85	4.66	22.47	18.52	57.28
Darbhanga	0.54	0.73	34.73	23.37	8.03	50.43
Gaya	–1.57	–0.53	–0.65	13.84	22.17	47.41
Muzaffarpur	0.32	0.99	22.19	21.98	11.37	47.40
Palamau	–1.07	–0.17	–0.90	15.06	24.31	49.59
Patna	–0.64	1.12	–3.34	7.73	21.54	42.29
Purnea	0.13	–0.89	76.93	44.88	16.54	49.33
Ranchi	0.21	1.51	13.42	14.21	25.12	37.74
Santhal Parganas	–1.60	–2.39	–4.72	25.06	–5.22	37.20
Saran	–0.08	4.53	5.97	22.40	16.37	50.32
Shahabad	0.08	3.32	1.71	21.59	32.92	43.85
Singhbhum	1.24	0.48	–2.16	19.43	–7.63	15.99
Dhanbad+Hazaribagh	–2.21	–3.77	–9.66	4.25	4.46	32.34
Monghyr+Saharsa	–0.09	–0.73	2.87	37.49	14.42	47.63

Source : Calculated by the Author.

TABLE 6.6: **Results of correlation analysis (area)**

Independent variables	Coefficient of correlation (*r*)	Coefficient determination (r^2)	Per cent explained variation ($r^2 \times 100$)
Irrigated area	0.301	0.0906	09.06
Fertilisers	0.383	0.1466	14.66
Tractors	0.030	0.0009	00.09
Pumping Sets	0.126	0.0158	01.58

Source Calculated by the Author.

In the case of the relationship between production (in terms of monetary value) and the four associated independent variables mentioned earlier, the "*r*" values are presented in Table 6.7. Again all four of the hypotheses have not been supported by the results of the correlation analysis. The "*r*" values have been found to be 0.00 with irrigated area, 0.07 with fertilisers, 0.50 with mechanisation — tractors — and 0.40 with pumping sets. Thus these values are very low and have been found to be not significant 0.5 probability level. However, the total variation in the dependent variable has been explained up to 42 per cent, which is much larger than in the previous case. It is also clear from Table 6.7 that the association between production (monetary value) and mechanisation is not so weak. In other words, there seems some impact of mechanisation—tractors and pumping sets—on the increase in the monetary value of production in the State.

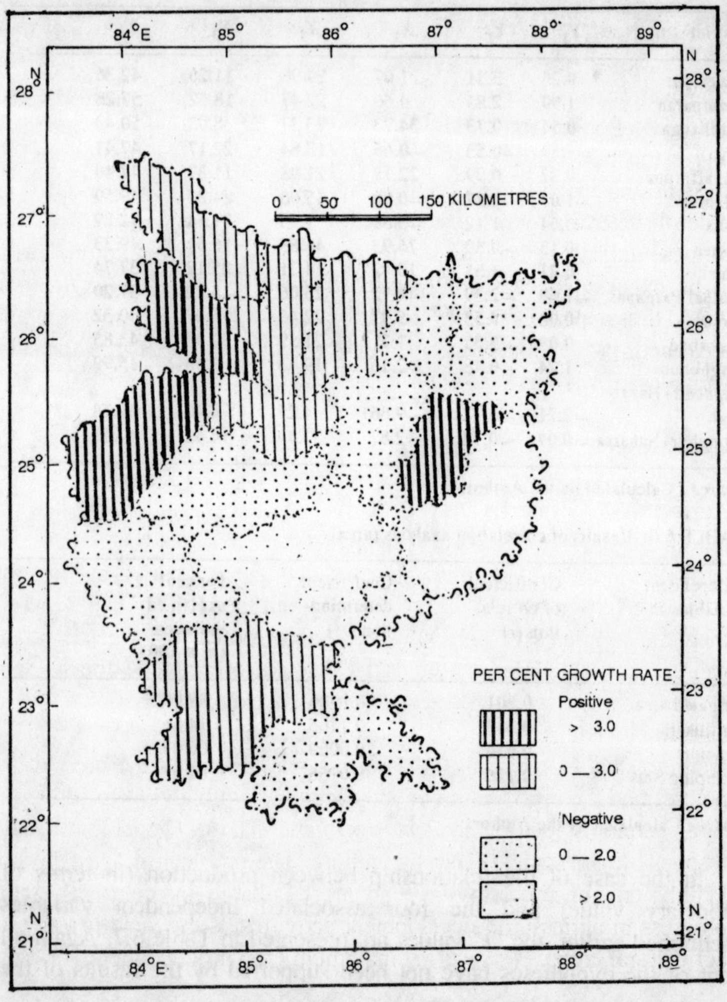

FIG. 6.3: Levels of agricultural growth — 1963-64 to 1971-72 (monetary value).

TABLE 6.7: Results of correlation analysis (monetary value)

Independent variables	Coefficient of correlation (r)	Coefficient determination (r^2)	Per cent explained variation ($r^2 \times 100$)
Irrigated area	0.004	0.0000	00.00
Fertilisers	0.079	0.0062	00.62
Tractors	0.502	0.2520	25.20
Pumping sets	0.403	0.1624	16.24

Source : Calculated by the Author.

Admittedly, these conclusions are based on a small range of data for 15 units (districts) only and as such have limited relevance. However, in view of the low values of correlation achieved, they give a just indication of the conditions prevailing in the region. The linear regression between area and production with associated independent variables has been shown on the scatter diagram (Figs. 6.4 and 6.5) in order to determine whether the relationship is in fact linear. Inspection of these diagrams reveals a significant departure from linearity. This is to be expected because of low "r" values prevalent among them. It would thus appear from the foregoing discussion that the growth rate in area and production has not been explained by the independent variables selected in this study.

Conclusion

Agriculturally, Bihar is very insecure and still continues to be at the mercy of the weather. The planners often attribute failure on the agricultural front to unfavourable weather conditions which have an adverse effect on agricultural production, but the problem is not merely that; the problems of agriculture are deep-rooted and weather alone cannot be blamed for its failure. Other factors include the tenancy system, small and fragmented holdings, uneven distribution of land, traditional equipment and technology, lack of proper irrigation facilities and dearth of capital. To this may be added the present-day politics and the introduction of social changes with a view to lifting the rural economy out of the rut overnight. These factors have been responsible for stagnation in agriculture. The problem of agriculture is mainly one of productivity. The most suitable indicator, assessing progress in agricultural development activities, is the yield per acre of agricultural crops. For securing maximum crop production, the best use of available land has to be made and the latest methods of crop husbandry have to be adopted.

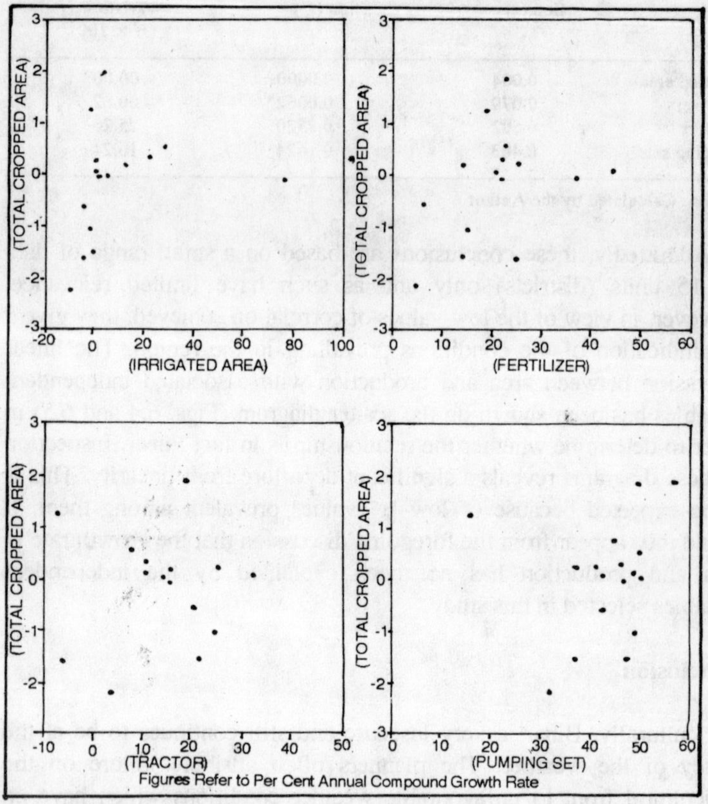

FIG. 6.4: Relationship between Total Cropped Area and Selected Agricultural Innovations in Bihar (1963-64 to 1971-72).

To conclude the following inferences may be drawn from the analysis of agricultural growth in Bihar : (i) the agricultural growth rates appear to represent an extremely complex phenomenon, particularly to the extent that they vary so significantly throughout the State; (ii) the study suggests that growth rate in production converted into monetary value is better explained than the variation in area with independent variables; (iii) independent variables like irrigation and fertilisers are apparently very much associated with increasing the productivity of, major crops, but in fact do not explain altogether the total variation. A cropwise calculation of "*r*" values may give a better result; (iv) to a geographer interested in explaining this areal variation in the levels of growth, the study appears to indicate that a larger number of other variables concerned with agriculture, for example, number of farmers, level of literacy, seeds, transportation and livestock will have to be considered; (v) the data on which the present analysis is based can be improved upon, since only Government sources have been considered to far; and (vi) the low value of correlation has been obtained due to the selection of district as a unit area; the smaller areal units might have proved more satisfactory to work with, as the spatial aggregation of data at the district level has concealed its reality.

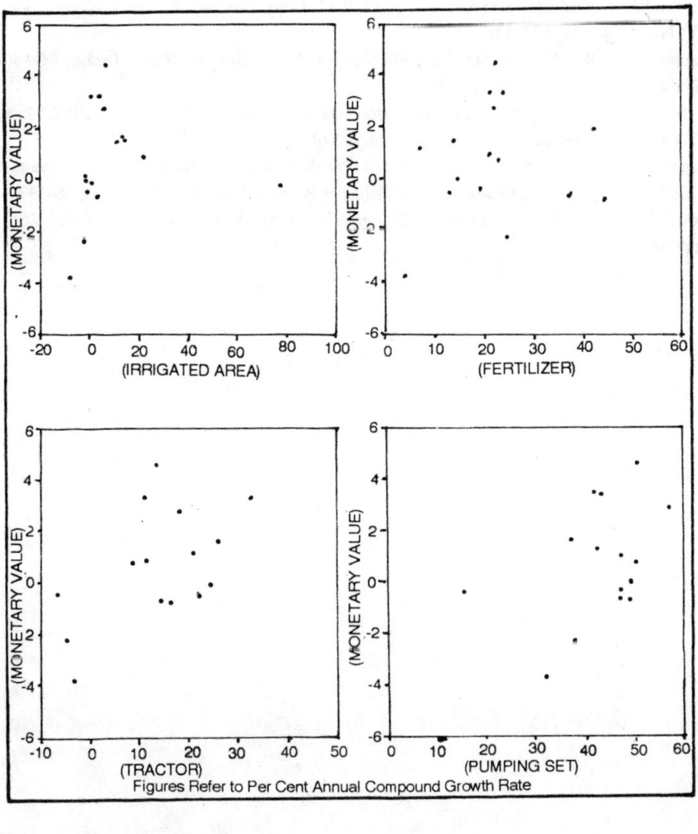

FIG. 6.5: **Relationship between Monetary Value and Selected Agricultural Innovations in Bihar (1963-64 to 1971-72).**

REFERENCES

Baruah, U., 1979 : "Role of High Yielding Varieties in Agricultural Development in Punjab and Haryana", in Mohammad, A. (Ed.), *Dynamics of Agricultural Development,* pp. 17-38.

Bhalla, S., 1977 : "Agricultural Growth : Role of Institution and Infra-Structural Factors", *Economic and Political Weekly,* Vol. XII, No. 45, pp. 1898-1904.

Huston, A., 1968 : "Variation in Agricultural Growth and Output between and within Regions of India", *Asian Survey.*

Khusro, A. M. (Ed.), 1968 : "Readings in Agricultural Development", Allied Publishers, New Delhi, p. 97.

Minhas, B. S. and Vaidyanathan, A., 1965 : "Growth of Crop Output in India 1951-54 to 1958-61 : An Analysis of Component Elements", *Indian Journal of Agricultural Economics,* Vol. 17, pp. 230-47.

Nath, V., 1970 : "Agricultural Growth in 1970's", *Economic and Political Weekly,* Vol. V, No. 52, pp. A. 134-44.

Ramachandran, R., 1975 : *Spatial Diffusion of Innovation in Rural India,* Mysore University Press, Mysore, pp. 38-43.

Rao, S. K., 1971 : "Inter-Regional Variations in Agricultural Growth 1952-53 to 1964-65", *Economic and Political Weekly,* Vol. VI, No. 27, pp. 1333-48.

Shangari, K. K. and Gupta, D. P., 1978 : "Growth of Agricultural Output in Haryana : 1952-53 to 1974-75", *Economic and Political Weekly,* Vol. XIII, No. 39, pp. 84-87.

Tarrant, J. R., 1974 : *Agricultural Geography,* John Wiley & Sons, New York, pp. 14-16.

7

AGRICULTURAL TRENDS IN WESTERN UTTAR PRADESH

RAGHUBIR CHAND AND KAILASH CHANDRA JOSHI

THE STUDY of agricultural trends of a given area forms the basis for planning and taking appropriate measures to raise agricultural production for meeting ever increasing demand of food. It is also essential for evaluating and improving the overall economic levels. Particularly after the advent of the new technology in agriculture, tremendous efforts have been made in India for increasing total supplies of food grains. Nevertheless, the performance of growth is quite uneven from one crop and region to another. Considering that western U.P. has witnessed a conspicious upheaval in its agriculture during last two decades, the present paper seeks to analyse the agro-economic trends of Western U.P. from 1965-66 to 1985-86. The study aims :

(i) To examine the overall changes in the broad landuse categories.
(ii) To interpret the changes in area, production and yield of crops during the period.
(iii) To determine the impact of irrigation, fertiliser and other technological inputs in the production levels.

The Study Area

The study area extends from 26°22′ N to 30°25′ North Lat. and 77°4′ E to 80°26′ East long covering an area of 82,192 sq km with a population of 3,93,49,930 persons as per 1981 census. Administratively, the region constitutes Rohilkhand, Meerut and Agra divisions; and Farrukhabad and Etawah districts of Allahabad division encompassing therein a total of 19 districts.

Physically, it forms the part of upper Ganga plain filled by the fine alluvium brought down by rivers Yamuna, Ganga, Sharda and their tributaries.

The Western Uttar Pradesh is essentially an agricultural region where about 68 per cent of the population rests upon agriculture for their livelihood. Although the agricultural conditions differ from western to eastern part, the area is endowed with favourable physical conditions in general. The region covers only 27.92 per cent of the total geographical area of the state where as it accounts for 35.50 per cent of the total population. It accounts for 36.9 per cent of the total cropland area, shares as much as 61.5 per cent of the gross agricultural production and almost half (49.2%) of the gross irrigated area. It also concentrates 60.66 per cent of the total number of tractors, 51.95 per cent of the irrigational pump sets, 38.48 per cent of the area under high yielding hybrid seeds and 45.05 per cent of the total fertiliser consumption of the state (Thakur, 1983). The average size of holding in the region is about 1.20 ha, while it is 1.01 hectare in the state. It is obvious, therefore, the present study would highlight the pattern of agricultural growth in a green revolution region.

Methodology

The basic data for the present study have been obtained from the agricultural statistics of U.P. from 1965-66 to 1985-86 in a year-wise sequence. The period of study is so selected as to capture the situation immediate after green revolution and the relatively recent patterns. Almost all main crops have been included namely paddy, wheat, maize, barley, pulses, oil seeds, sugarcane, potato and tobacco. Besides computing the percentage change in area and production, three years moving averages of area and output have also been worked out. In order to measure the variability patterns, trend line has been fitted with a regression line of the form $YC = a + bx'$. The regression line makes it possible to predict the exact change in the production level in various time points. Similarly the three years moving averages have been also plotted in the time series graph so as to clear the trend of change. The correlation coefficient is worked out and tested for significance by '*t*' test.

Landuse Pattern

Landuse is an important aspect of geographical studies and the progress of an area can be measured to a certain degree by the way in which its land is used and maintained (Rao, 1986). The landuse pattern is determined by several interrelated factors like the environmental, socio-economic and also the scientific management of the landuse itself. Table 7.1 shows the area under different categories of landuse as percentage to total reported area. The percentage of area under forest is very low which essentially means intensive use of land for cultivation. The area covered by forest is merely 4.67 per cent in 1985-86, while it was 4.84 per cent of the total reported area in 1965-66. The barren and uncultivable area has also declined from 4.53 per cent in 1965-66 to 3.32 per cent in 1985-86, while the land put to non agricultural uses including area under roads, railways and settlements increased significantly from 7.81 per cent in 1965-66 to 9.27 per cent in 1985-86.

The percentage share of uncultivated area other than fallows also declined marginally under which the sub-category of miscellaneous tree crops declined from 1.49 per cent to 0.65 per cent, while the category of pasture land registered negligible but positive change from 0.26 per cent in 1965-66 to 0.28 per cent in 1985-86. However there has been a substantial change in the extent of culturable waste land which declined from 4.12 per cent in 1965-66 to 2.18 per cent in 1985-86. Likewise the current fallow has also undergone changes which is noted as 3.70 per cent in 1965-66 and 3.11 per cent in 1985-86. Albeit other fallow land is increased from 1.39 to 2.59 per cent during the period. Another major change has been observed in the percentage share of net sown area which is increased from 71.86 per cent to 73.93 per cent. The increase in net cultivated area can be explained by the decline in the forest area, barren land, miscellaneous tree crops and culturable waste land. The double cropped area has been also increased tremendously registering 54.16 per cent of net sown area in 1985-86 from 28.20 per cent in 1965-66. The variation is worked out to be 25.90 per cent during these two decades. Table 7.1 brings out the year-wise details which broadly reflect a consistent pattern of growth in cultivated area which is essential for agricultural development. The highest area under cultivation was recorded in 1981-82.

TABLE 7.1: Landuse pattern of Western U.P. (Percentage of reported area) from 1965-66 to 1985-86

Year	1	2		3		4	5		6	7
		I	II	I	II		I	II		
1965-66	4.84	7.81	4.53	0.26	1.49	4.12	3.70	1.39	71.86	28.20
1966-67	4.98	7.91	4.48	0.26	1.47	3.98	3.37	1.41	72.14	31.21
1967-68	4.74	7.95	4.48	0.28	1.46	3.83	3.30	1.37	72.59	35.93
1968-69	4.74	8.03	4.35	0.29	1.47	3.88	3.23	1.50	72.51	34.14
1969-70	4.81	8.10	4.29	0.32	1.45	3.81	3.10	1.50	72.62	35.52
1970-71	4.50	8.25	4.19	0.32	1.43	3.80	3.46	1.59	72.48	39.40
1971-72*	4.58	8.22	4.18	0.31	1.43	3.75	3.57	1.57	72.39	35.66
1972-73	4.68	8.26	4.17	0.32	1.43	3.38	3.75	1.45	72.36	39.05
1973-74*	4.79	8.62	3.59	0.31	0.90	4.26	3.69	1.65	72.19	41.88
1974-75	4.67	8.46	4.06	0.31	1.40	3.43	3.72	1.66	72.29	39.00
1975-76	4.66	8.57	3.98	0.32	1.39	3.31	3.64	1.81	72.32	41.46
1976-77	4.65	8.62	4.09	0.32	1.30	3.08	3.16	1.69	73.09	40.99
1977-78	4.64	8.64	3.78	0.31	1.28	2.90	3.33	1.71	73.41	39.97
1978-79	4.64	8.74	3.67	0.30	0.83	2.82	3.19	1.82	73.99	46.93
1979-80	4.74	9.06	3.79	0.30	0.77	2.45	3.18	2.13	73.58	46.99
1980-81	4.70	8.93	3.57	0.29	0.75	2.20	3.60	2.36	73.60	50.71
1981-82	4.67	9.02	3.46	0.28	0.70	2.23	3.14	2.34	74.16	48.97
1982-83	4.66	9.02	3.46	0.29	0.64	2.22	3.32	2.32	74.07	50.02
1983-84	4.69	9.09	3.39	0.29	0.63	2.18	3.20	2.43	74.10	52.87
1984-85	4.68	9.17	3.42	0.28	0.60	2.15	3.20	2.52	73.98	53.88
1985-86	4.67	9.27	3.32	0.28	0.65	2.18	3.11	2.59	73.93	54.16

* Estimated.

Source: Agricultural Statistics of U.P. for the year 1965-66 to 1985-86, U.P. Lucknow.

1. Forest.
2. Land not available for cultivation —
   - I. Land put to non-agricultural uses,
   - II. Barren and uncultivable land.
3. Other uncultivated land (excluding fallow land) —
   - I. Permanent pasture and other grazing land,
   - II. Land under miscellaneous tree crops and groves.
4. Culturable waste.
5. Fallow land —
   - I. Current fallow,
   - II. Other fallow.
6. Net area sown.
7. Area sown more than once.

Cropping Pattern

The study of cropping pattern involves the description of area under various crops and change there in overtime gives a clue to understand the option preferred by farmers. The cropping pattern differs from region to region both in space and time, and it is largely governed by the physical, cultural and technological factors (Giri & Misra, 1983). Table 7.2 presents area under some important crops as percentage to the gross cropped area. It reveals that wheat predominates over all other crops and the magnitude of growth rate of area of wheat is also higher. In 1965-66, it covered 22.87 per cent, while in 1985-86 it rose to 32.97 per cent registering an increase of 10.10 per cent of gross cropped area. The shift in cropping pattern is very irregular due to the interaction of physical environment on the one hand and the responsible socio-economic factors on the other (Singh & Singh, 1970). It is significant to note that area under pulses and sugarcane was higher than paddy in 1965-66, while in 1985-86, paddy ranks second followed by sugarcane, pulses, maize, oilseeds, barley, potato and tobacco. Thus the area under paddy has also increased considerably from 10.04 to 12.48 per cent. Other crops which registered marginal increase in the area are oil seeds (1.01%), potato (0.74%), and maize (0.14%). On the contrary, pulses (–10.96), sugarcane (–0.80%) and barley (–1.25%) have registered a decline over the years. Viewing the growth in the area of wheat, it is found that it continued to raise year after year with the only exception of 1973-74 and 1974-75 years when it declined by 1.0 per cent. The contribution to the growth of wheat area was made mainly by decrease in the area under pulses which also followed a continuous decline. This consistent behaviour of growth is not observed in the case of paddy which has shown much fluctuations. It reflected a slow growth which can be explained by the limitations posed by the fixity of land. The area under maize grew upto 9.88 per cent in 1971-72 and further declined. These results aptly suggest that the area shift from coarse cereal and pulses to mainly wheat and paddy is due to the new package programmes widely introduced in wheat and rice production. It has inversely affected the sugarcane which shows decline in area once sharing a considerable percentage of state's income. These variations have their genesis in the human choice to grow crops, economic monetary and the infrastructural built up of the area gradually after the green revolution.

TABLE 7.2: Western U.P. percentage of area under principal crops to gross cropped area (1965-66 to 1985-86)

Years	Paddy	Maize	Wheat	Barley	Pulses	Oil-seeds	Sugar-cane	Potato	Toba-cco
1965-66	10.04	6.73	22.87	3.81	18.88	3.72	11.18	0.79	0.10
1966-67	10.73	7.18	24.39	3.85	18.55	3.97	8.39	0.70	0.11
1967-68	11.03	7.97	26.53	3.95	17.04	4.02	9.73	0.70	0.11
1968-69	11.31	8.09	27.67	3.20	15.19	3.14	8.06	0.96	0.10
1969-70	10.91	8.48	28.85	3.19	13.70	3.13	9.87	0.76	0.10
1970-71	10.52	8.77	31.61	2.59	11.51	3.31	9.29	0.82	0.08
1971-72*	11.73	9.88	33.57	2.69	11.11	4.07	9.06	0.94	0.09
1972-73	10.12	8.99	33.90	2.50	10.28	3.17	9.36	1.04	0.09
1973-74	11.13	8.98	31.66	3.00	10.11	3.51	10.59	0.97	0.07
1974-75	10.47	8.14	30.74	4.23	8.98	5.04	11.20	1.15	0.05
1975-76	11.11	8.46	30.60	4.07	8.76	4.50	10.59	0.99	0.09
1976-77	11.22	7.97	30.96	3.62	8.41	4.56	11.83	0.99	0.09
1977-78	11.95	6.33	32.47	2.90	7.49	4.35	12.64	1.06	0.08
1978-79	12.89	5.93	33.65	2.60	8.17	3.56	11.73	1.46	0.08
1979-80	12.46	6.31	34.63	2.59	7.71	2.67	10.04	1.27	0.10
1980-81	12.29	6.81	34.19	2.70	7.34	3.75	9.86	1.36	0.14
1981-82	11.58	6.52	31.95	2.59	7.99	4.83	11.98	1.52	0.10
1982-83	10.52	5.83	34.57	2.28	7.49	4.44	12.64	1.39	0.10
1983-84	11.79	5.87	35.40	2.33	6.33	3.33	11.80	1.56	0.13
1984-85	12.33	6.62	34.25	2.20	6.42	4.67	10.79	1.69	0.15
1985-86	12.48	6.87	32.97	2.56	7.92	4.73	10.34	1.53	0.14

* Estimated.

Production Variability

Table 7.3 shows percentage change in the production as compared to the previous year. Wheat is the major crop of the area. The output of wheat increased (4.45 times) from 17,18,219 tonnes in 1965-66 to 76,54,427 tonnes in the year 1985-86. It came down between the period 1970-71 and 1974-75 from 40,57,107 tonnes to 34,46,741 tonnes. The maximum decline (–32.96%) was registered in 1973-74 followed by the year 1979-80 (–11.27%), 1981-82 (–5.29%) and 1984-85 (–3.21%). The production of paddy increased slightly higher than wheat (4.6 times) form 4,61,285 tonnes in 1965-66 to 21,63,127 tonnes in 1985-86. When the relative trend is observed, three time periods provided the similar feature of decline for both wheat and paddy namely 1972-73, 1979-80 and 1981-82. The year 1980-81 is very important for the paddy from the view point of production variability being highest 104.06 per cent. The temporal variation is almost similar for both wheat and paddy as evident from a common graph (Fig. 7.1). The production of maize increased (1.72 times) from 5,21,053 tonnes to 8,99,463 tonnes and the similar trend of growth (1.65 times) is observed for barley which increased from 2,65,929 tonnes to 4,39,527 tonnes during the study period. Pulses, however, witnessed a marked decline (1.74 times) from 12,86,962 tonnes to 7,39,746 tonnes.

Oilseeds, potato, tobacco and sugarcane have also registered 1.37, 2.50, 3.42 and 1.26 per cent increase during the period. These figures finally reveal that though the output of the major crops rose due to the impact of new agricultural technologies, there has been considerable variation in annual production which emphasises the importance of climatic factors as a major determinants of agricultural production. In fact annual production changed in the same direction as the annual rainfalls of the years under comparison.

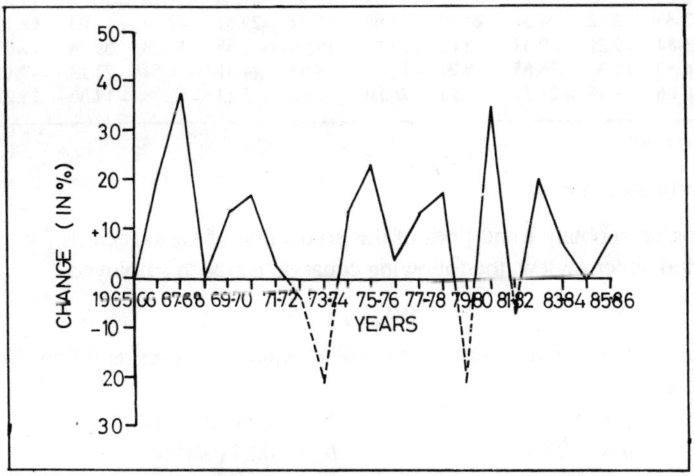

FIG. 7.1: **Western U.P. temporal variation in production of padddy (in terms of rice) and wheat (1965-66 to 1985-86).**

TABLE 7.3: **Western U.P. percentage change in the production of major crops as compared to previous year**

Years	Paddy	Maize	Wheat	Barley	Pulses	Oil-seeds	Potato	Toba-cco	Sugar-cane
1965-66	—	—	—	—	—	—	—	—	—
1966-67	23.53	–5.23	19.44	7.66	–38.97	–11.41	—	525.67	–29.79
1967-68	37.09	5.42	38.13	38.16	57.54	26.59	—	—	3.33
1968-69	–12.72	20.01	3.59	–44.29	–15.67	–40.67	—	—	–4.80
1969-70	18.27	–3.45	11.82	39.85	4.37	13.03	–31.30	30.89	19.48
1970-71	–1.84	71.83	23.54	–10.55	–19.53	3.29	27.22	–20.15	–9.32
1971-72*	20.94	–32.36	–1.95	2.0	–2.86	–6.67	12.81	2.93	–2.99
1972-73	–21.15	0.01	–0.34	–8.99	–9.94	25.47	–1.65	—	10.43
1973-74*	39.70	2.89	–32.96	–4.09	–42.76	0.58	5.94	—	14.78
1974-75	–28.44	43.00	29.69	85.82	47.36	40.46	33.41	37.43	1.72
1975-76	42.28	47.83	18.34	–6.38	14.78	–16.54	10.36	80.80	–4.15
1976-77	2.86	8.18	4.53	–14.99	–1.05	–18.32	—	—	17.79
1977-78	27.97	–40.77	9.68	–16.33	–13.60	–0.55	—	—	16.67
1978-79	19.92	–14.16	16.93	0.22	–8.15	–19.95	50.92	1.59	–19.84
1979-80	–54.45	56.79	–11.27	–18.73	–31.48	–25.95	–25.17	34.73	–15.50
1980-81	104.06	10.26	24.46	53.80	47.75	122.92	51.54	51.39	28.86
1981-82	–13.83	9.23	–5.79	–19.14	–8.98	49.58	–1.18	–23.92	12.01
1982-83	2.12	–20.31	24.97	8.67	15.22	–29.53	–7.66	3.03	10.10
1983-84	29.21	29.63	3.05	13.87	–19.90	–2.88	45.18	39.28	–4.01
1984-85	11.36	76.65	–3.21	–10.33	8.83	44.51	–4.54	21.13	–7.11
1985-86	8.45	–21.59	7.94	28.80	37.12	–5.13	–52.08	–14.36	–15.21

* Estimated.

Production Trend

In order to obtain trend lines of the production of various crops for the period under review, the following equation has been employed:

$$YC = a + bx$$

The trend as worked out with the above equation comes as follows for various crops:

$a = 52.10$, $b = 1.89$ (maize)
$a = 26.59$, $b = 0.87$ (barley)
$a = 22.42$, $b = 0.42$ (oilseeds)
$a = 171.82$, $b = 29.68$ (wheat)
$a = 46.13$, $b = 8.51$ (paddy)
$a = 128.77$, $b = -2.74$ (pulses)

These trends values have been plotted in Figs. 7.2-A and 7.2-B so as to make an easy comparison, Fig. 7.2-A suggests that production of wheat increased rapidly as compared to rice which increased but moderately. On the other hand pulses showed rapid decline. The graph plotted for barley, maize and oilseeds shows a slow growth.

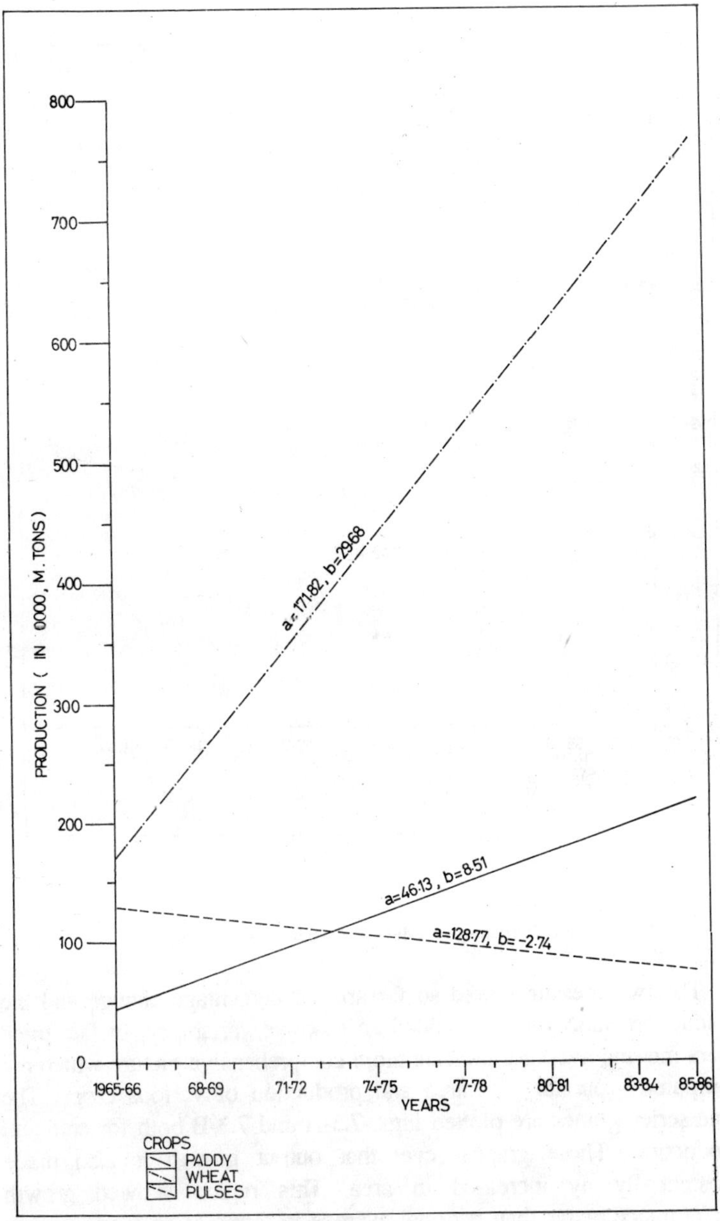

FIG. 7.2-A: Western U.P. trends of production (1965-66 to 1985-86).

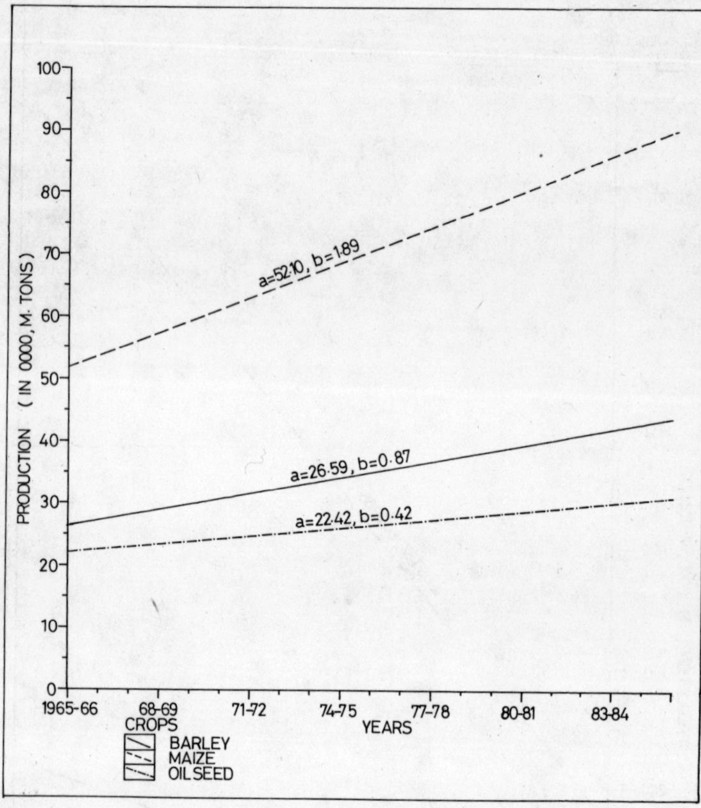

FIG. 7.2-B: Western U.P. trends of production (1965-66 to 1985-86).

The two measures used so far are the percentage change and the production trend which provided an absolute variability. In fact three years moving averages indicate more comprehensive picture which are computed separately for area and production of various crops. The time series graphs are plotted Figs. 7.3-A and 7.3-B both for area and production. These graphs revel that output growth is also made substantially by increased in area. This region showed growth performance better than national average in terms of area and output. The correlation coefficient worked out between area and production for these crops further confirm the contribution of area in raising the production level.

TABLE 7.4: Relation between area and production worked out from three years moving average data

Crops	'Y' value	't' value
Paddy	0.88	8.0759**
Wheat	0.91	9.5671**
Maize	0.42	2.0173*
Barley	0.49	2.4502*
Sugarcane	0.94	12.0096**
Oilseeds	0.82	6.2448**
Pulses	0.83	6.4864**
Potato	0.93	11.0289**

* Significant at 0.05 level, and
** Significant at 0.01 level.

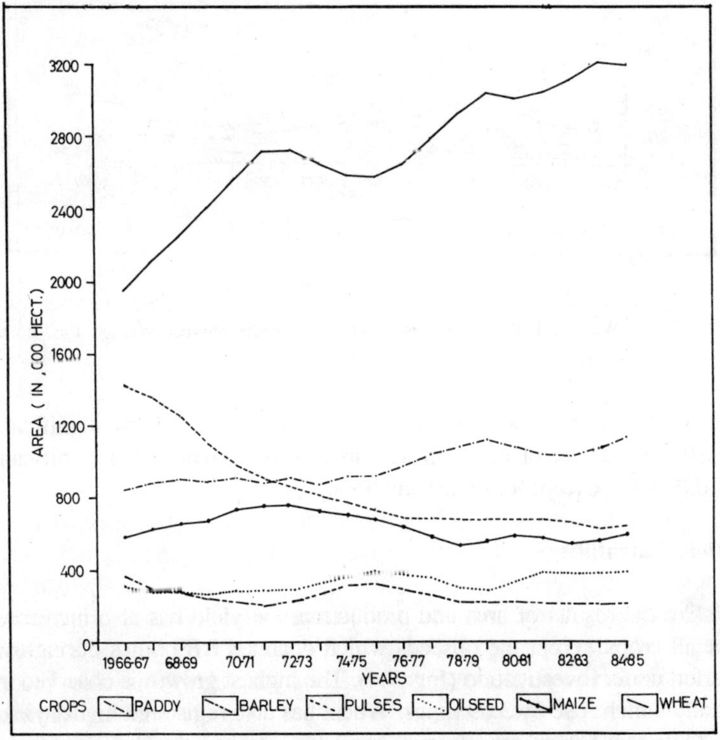

FIG. 7.3-A: Western U.P. three years moving average (1965-66 to 1985-86).

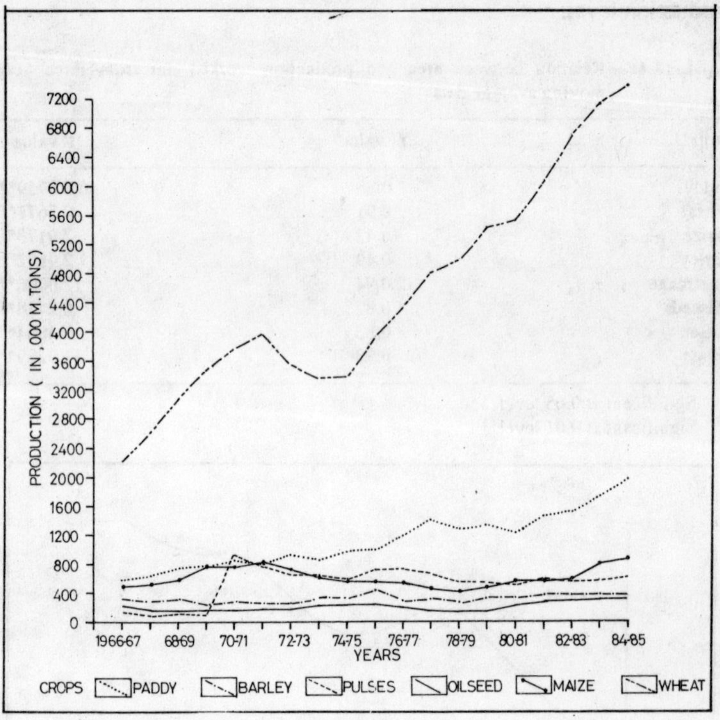

FIG. 7.3-B: Western U.P. time series graph (three years moving average 1965-66 to 1985-86).

As expected the production showed a positive and highly significant coefficient of correlation with the area in all the cases and significant at 0.01 level except for maize and barley.

Yield Variability

Unlike the results of area and production, the yield has also increased for all crops except the oilseeds which declined 0.89 times during the period under investigation (Fig. 7.4). The highest growth is observed in paddy which rose by 3.09 times. Wheat has also registered higher yield (2.54 times) which rose from 979.52 kg. per hectare in 1965-66 to 2428.02 kg. hectare in 1985-86. There is also a definite upward trend in case of barley (2.01 times), maize (1.38 times), sugarcane (1.12 times) and potato (1.05 times). Table 7.5 presents yield of different crops. The growth behaviour of yield is almost synonymous to the growth in production levels. However, in case of pulses, the area and production tended to decline but the yield per hectare is increased about 0.89 times. On the contrary yield of oilseeds such as rapeseeds and mustard declined by 0.89 times, while area and production have increased during the period under investigation.

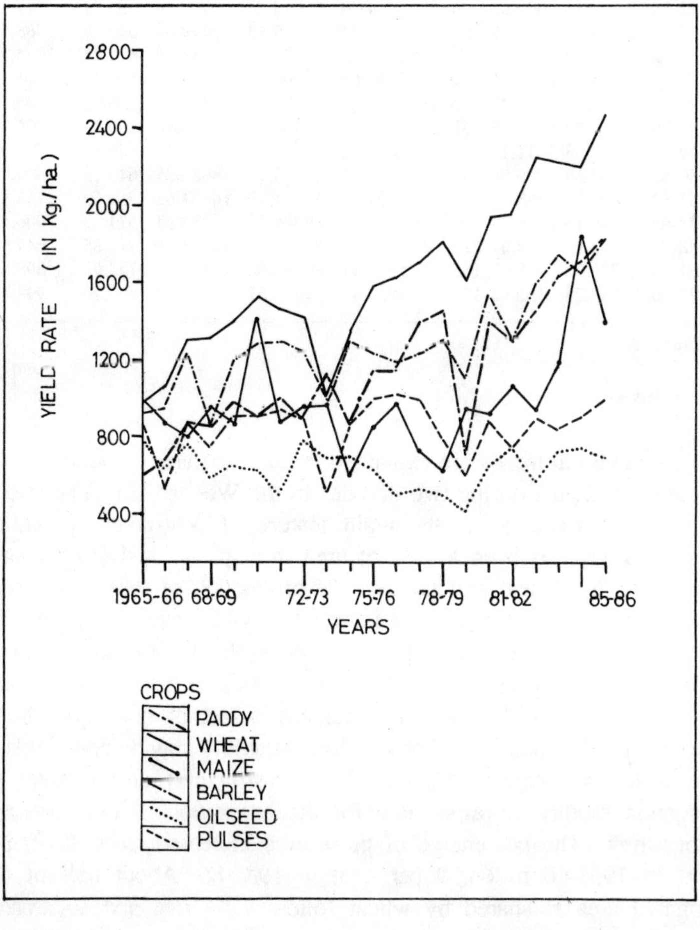

FIG. 7.4: Western U.P. average yield (1965-66).

TABLE 7.5: Western U.P. yield per ha. in kg. for different crops (1965-66 to 1985-86)

Years	Paddy	Wheat	Maize	Barley	Sugarcane	Potato	Oilseeds	Pulses
1965-66	599.33	979.52	1009.67	909.67	39287.90	9349.71	786.71	888.58
1966-67	673.59	1067.10	871.92	942.85	35768.06	—	634.31	537.00
1967-68	864.11	1303.34	796.59	1220.02	39526.34	—	762.88	885.68
1968-69	747.30	1315.42	957.25	854.38	42965.77	9571.16	589.39	851.21
1969-70	905.27	1394.24	782.09	1184.90	44541.77	8204.33	658.93	973.88
1970-71	902.50	1538.58	1416.80	1273.64	42001.22	9398.05	631.36	912.78
1971-72*	1004.45	1458.91	873.57	1285.41	42904.69	9589.48	491.41	943.71
1972-73	905.13	1419.49	946.80	1240.30	45212.75	8386.80	780.69	905.56
1973-74*	1129.14	1000.09	958.01	974.80	45013.23	9368.70	795.71	517.10
1974-75	896.47	1352.57	610.27	1301.94	43825.13	10629.89	690.06	868.52
1975-76	1145.48	1579.93	852.28	1242.10	43637.46	13434.09	632.74	1004.75
1976-77	1159.28	1621.11	972.68	1180.65	45728.91	—	506.96	1028.54
1977-78	1396.86	1700.78	727.66	1236.31	50064.79	15314.60	530.91	1000.79
1978-79	1467.82	1814.50	629.91	1307.02	39933.57	15752.80	490.43	796.38
1979-80	712.58	1611.16	956.97	1099.82	40590.94	13972.69	426.61	595.93
1980-81	1407.67	1939.96	933.20	1547.76	50914.42	18883.25	643.13	882.78
1981-82	1292.62	1963.98	1069.99	1307.90	47131.93	16834.62	750.59	741.73
1982-83	1446.16	2256.34	948.39	1606.54	48926.21	16883.93	571.97	906.47
1983-84	1633.92	2226.07	1196.40	1755.28	49292.69	21434.46	726.85	841.59
1984-85	1730.92	2215.97	1853.26	1662.11	49853.62	18789.61	745.97	898.87
1985-86	1853.28	2482.02	1399.30	1837.45	44037.51	9896.57	697.79	998.40

* Estimated.

Conclusion

The agricultural trends are explained through the area, production and yield components during two decades in the Western U.P. The rise in all three components is the main feature of post-green revolution period. There has been a shift of area from pulses and other course cereals to the wheat and paddy. The production of these two crops grew rapidly with the marked increase in yield level. However, the yield per hectare for all crops increased with the only exception of oilseeds. It, therefore, offset the effect of area in the production and indicated towards the more pronouncing effect of new agricultural technology. The analysis of production variability may be concluded in the following manner. Firstly, the infrastructural development of irrigation facility is responsible for the adoptation of new package programme. The percentage of gross area increased from 40.70 per cent in 1965-66 to 68.09 per cent in 1985-86. About half of the irrigated area is shared by wheat followed by rice and sugarcane. Irrigation is necessary not only for the extension of the cultivated area but it also gave rise in the percentage share of area under double cropping. The means of irrigation also changed vehemently in 1965-66. Canals accounted for 44.89 per cent of irrigated area. While the area irrigated by other wells was 29.78 per cent and 23.07 per cent by tubewells, 10.09 per cent by ponds and lakes and 1.17 per cent by other means. In 1985-86, Tube wells emerged to be the primary source contributing 64.56 per cent of irrigated area. The share of canals, other wells and ponds declined to 25.77, 6.61 and 0.26 per cent with a marginal increase in other source, i.e., 2.80 per cent. The large scale introduction of electric pumps and oil engines is responsible for the increasing use of tube wells. This has consequently gave rise to the consumption of fertilisers during the same period. The total fertiliser used in 1962-65 was 31,828 tonnes which showed 27.79 times increase being 8,84,572 tonnes in 1985-86. The percentage of area under high yielding varieties to gross cropped area also increased from less than 30 per cent to more than 75 per cent during the same period. Western Uttar Pradesh has also witnessed remarkable rise in the number of agricultural implements particularly tractors, oil engines and electric pumps.

The above supporting informations indicate clearly that the level of agricultural productivity improved mainly due to the manifold increase in the modern agricultural implements and use of HYV and fertilisers. At the same time, the production fluctuated considerably due to inadequate rainfall and other environmental factors. The years of 1968-69, 1973-74, 1974-75, 1979-80 and 1981-82 have been, perhaps, the most difficult years when the output dropped significantly. This leads us to conclude that weather elements have still a bearing on the agriculture. However, the fosselfuel based agriculture which can be sustained with a higher economic cost has so far brought out higher per hectare out-turn in Western Uttar Pradesh.

REFERENCES

Giri, H. H. and Misra, Y. K., 1983 : Crop Diversification in Tehsil Hata, *The Indian Geographical Journal*, Vol. 58, No. 2, p. 57.

Rao, V. L. S. Prakash, 1986 : Landuse Survey of India, in P.S. Tewari (ed.), *Agricultural Geography*, Vol. VIII, Herotage Publishers, New Delhi, pp. 28-43.

Singh, K. N. and Singh, B., 1970 : Landuse, cropping pattern and their ranking in Shah Gunj Tehsil: A Geographical Analysis, *National Geographical Journal of India*, Vol. 16, No. 3-4, p. 223.

Thakur, M. C., 1983 : *Spatial Variation and Determinants of Agricultural Productivity in Western Uttar Pradesh 1962-65 to 1978-81*, M. Phil Dissertation, C.S.R.D., J.N.U., New Delhi, p. 139.

[illegible] The [illegible] of [illegible] [illegible] [illegible] [illegible] and [illegible] [illegible] [illegible] [illegible] and [illegible] [illegible] [illegible] [illegible] [illegible] and 7.6 per cent [illegible] [illegible] [illegible] 2.46 per cent [illegible] [illegible] [illegible] [illegible] [illegible] [illegible] [illegible] [illegible] and [illegible] percentage [illegible] [illegible] [illegible] [illegible] [illegible]

[illegible] [illegible] [illegible] [illegible] [illegible] [illegible] [illegible] [illegible] [illegible] [illegible] [illegible] [illegible] [illegible] [illegible] [illegible] the same [illegible] [illegible] [illegible] [illegible] [illegible]

REFERENCES

[illegible]

[illegible]

[illegible] (1970) [illegible] *Geographical Analysis* [illegible] *Geographical Journal of India* [illegible]

[illegible]

8

IMPACT OF REGIONAL DISPARITIES ON THE DEVELOPMENT OF AGRICULTURE IN BIJAPUR DISTRICT

G. V. TOGARSI, S. S. NAREGAL AND C. N. MATTIMANI

TOWARDS THE SOCIO-ECONOMIC upliftment of the rural sector is one of the basic themes of research now a days. Bijapur district, located in the northern plain of Karnataka plateau is one of the drought-prone districts of India. It has the largest geographical area (17,069 sq.km.) in the State consisting of 1,239 villages and 19 towns. The statistical analysis reveals that the district has good percentage of arable land i.e., 14,22,713 hectares (83.09 per cent of the total geographical area) in 1984-85, out of which 10.84 per cent is under irrigation. According to 1981 census, the population of the district is 24,01,782 and its density is 148 per sq.km. which is less than the State's average (191). The rate of literacy is 31.96 per cent and the urbanisation is 24.09 per cent. Of the total population (24,01,782), 41.60 per cent are workers and 58.40 per cent are non-workers. Out of working population the cultivators are 34.80 per cent, agricultural labourers 38.01 per cent, people engaged in household industries 6.03 per cent and other workers are 21.16 per cent. This indicates that the agriculture is the main occupation of people in the district. Being located in semi-arid tract of Northern Maidan, the district enjoys dry and healthy climate with an average annual rainfall of 554 mm. The average maximum temperature is 40.00° and minimum 14.8°C. Five rivers are passing through the district and the district has fertile black soils covering major portion and the remaining area is covered by shallow black and red soils. The occurrence of droughts at least twice in every five years has made this region economically retarded. The study of social, economic and demographic indicators, therefore, certainly explains that Bijapur district is under developed in agriculture, literacy and economic progress. Reducing the regional disparities in the levels of agricultural development is of major concern for any spatial planning. Development is a multi-dimensional phenomena which is governed by several factors. The modes of development like industrialisation, transport and communication, banking facilities, educational institutions, health centres, technological and institutional innovations in agriculture are the major factors contributing for the levels of agricultural development. However, in reality, it may not be possible to get uniform distribution of these infrastructural facilities in India. As a result, the regional disparities are bound to occur in a given time and space.

An attempt is made here to study the impact of regional disparities on the levels of development of agriculture (Fig. 8.1). The levels of agricultural development is studied by taking 16 indicators for the district through which the areas of low, medium and high development are noticed. During the study period (1984) Badami taluk comes in the very high development group. Mudhol, Jamkhandi, Bilgi, Bagalkot and Hungund are under medium development, Sindgi, Muddebihal and Bijapur are indicating low development. A large number of taluks come under medium development. Yet, about 45 per cent of the area in the district is still under low and very low groups.

Based upon the 16 indicators the level of agricultural development are identified. The regional disparities have been studied by taking 28 indicators (Fig. 8.2). These 28 indicators are grouped into social, economic and demographic factors. Bijapur taluk which is noticed as low development area in the agriculture development has come in the very high developed group (based on 28 indicators) which is mainly because of its location as district headquarter, transport and communication facilities, financial aid giving institutions, use of fertilisers and the development of irrigational facilities by wells. Similarly Jamkhandi, Hungund, Indi, Bagewadi and Badami get high development due to the increase in irrigational facilities, mechanisation, use of HYV seeds and application of modern devices in agricultural practice. These taluks have shown improvement over the levels of regional development except Badami. The study on regional disparity does not show the areas of low development. The cor-relation worked out between levels of agricultural development and the regional disparities shows a positive trend.

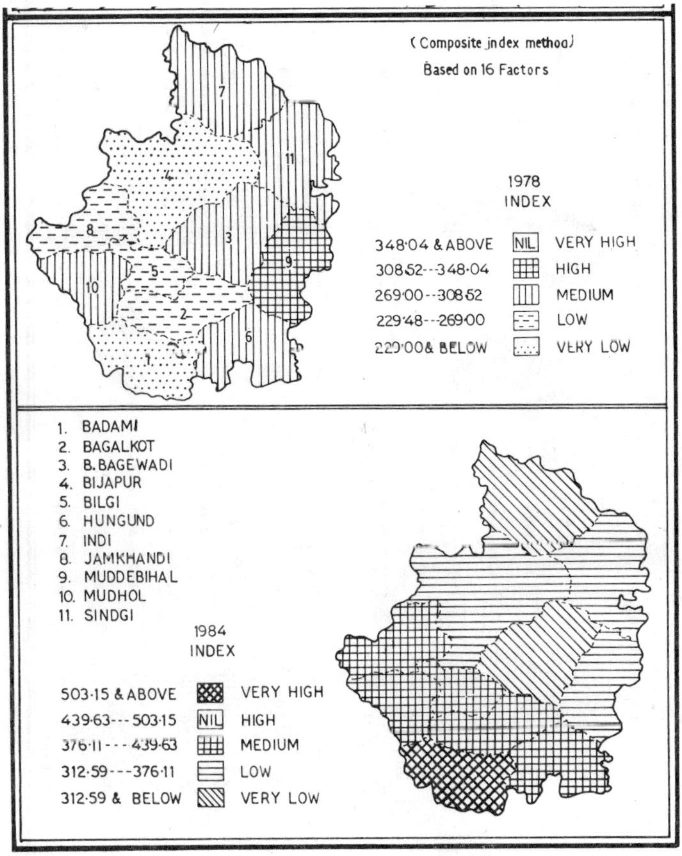

FIG. 8.1: Bijapur district: levels of agricultural development.

In order to correlate the level of agricultural development with the levels of regional disparity Spearman's rank order correlation has been calculated. The '*r*' value has been calculated for the three groups of indicators, i.e., social, economic and demographic indices of levels of agricultural development. The '*r*' value of agricultural development

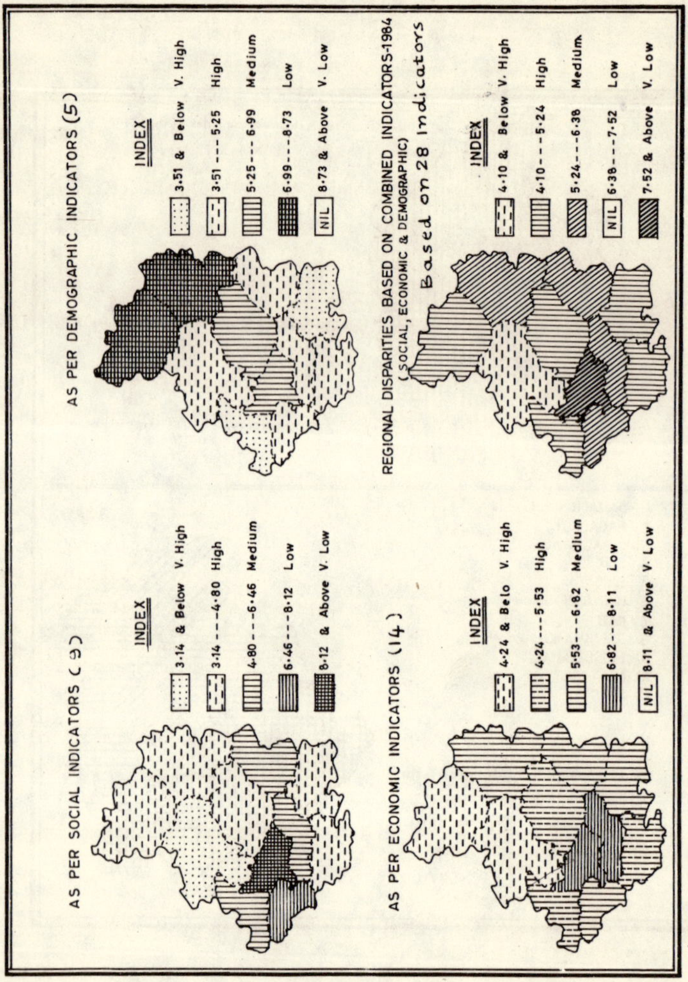

FIG. 8.2: Bijapur district: regional disparities: 1984.

has been found to be 94.30 for social indicator, 93.55 for economic indicator and 88.64 for demographic indicator. The combined index, therefore, is 91.77. This value (91.77) is very high and indicates maximum and significant correlationship. It means that the high level of regional development leads to the high development of agriculture and vice-versa. The above inter-relationship, therefore, shows that the regional development and agricultural development go hand in hand. Setback in any one of them leads to the setback of the other.

Objectives of the Study

1. To know the socio-economic and demographic structure in the rural sector of Bijapur district to reduce the regional disparities in the levels of agricultural development.
2. To test the level of agricultural development in relation to the existence of physical, cultural and socio-economic factors.
3. To regionalise the regional disparities in relation to the level of agricultural development.
4. To workout the trend of regional disparities in the district.

Methodology

In this study an attempt is made to find out the relationship between regional disparities and the levels of agricultural development of Bijapur district. To find out levels of agricultural development 16 factors are analysed with the help of composite index. Since all the indicators of development are not equally important, different, weights are assigned to different indicators by the method of *per cent proportional standardised mean*. That is to say the weight assigned to one indicator is calculated by using $\bar{X}/\sigma$ for each indicator, where $\bar{X}$ is the series of one particular indicator and is the standard deviation for the same series. This $\bar{X}/\sigma$ is the weight of each indicator. The formula for the composite index number used is .

$$C.I. = \frac{X_1 \; W_1 + X_2 \; W_2 \; --- \; X_{16} \; W_{16}}{W_1 + W_2 + W_3 + ---- \; W_{16}}$$

X = Raw score of any indicator of development

$$W_1 = \frac{\bar{X}}{\sigma} \text{ or } \frac{\text{Mean}}{\text{S.D.}}$$

In order to get the index of agricultural development the following 16 indicators have been selected after carefully examining their degree of importance in determining the agricultural development in Bijapur district :

1. Irrigated area as percentage to the net sown area.
2. Cropping intensity.
3. Agricultural productivity based on yield per hectare.
4. Degree of commercialisation as percentage of commercial cropped area to the total cropped area.
5. Number of tractors and irrigation pumpsets per 10,000 hectares of cropped area.
6. Livestock density per 10,000 hectares of agricultural area.
7. Net Sown Area as percentage to the geographical area.
8. Fertiliser consumption in tonnes.
9. Area under HYV as percentage to the net sown area.
10. Area sown more than once as percentage to the net sown area.
11. Per capita availability of cropped area to the total population.
12 Electrified villages as percentage to the total villages.
13. Rate of literacy as percentage to the total population.
14. Percentage owner cultivators as percentage to the total population.
15. Number of market centres.
16. Agricultural loan giving institutions.

Stratification of taluks according to the levels of agricultural development (The taluks given in the composite index have been divided into the following categories of levels of agricultural development with the help of mean and standard deviation).

1978

Levels of Development	Composite Index Range	No. of Taluks	Name of the Taluks
1	2	3	4
Very High	348.04 and above	Nil	Nil
High	308.52 to 348.04	2	Badami, Bijapur.
Medium	269.0 to 308.52	3	Jamkhandi, Bilgi, Bagalkot.
Low	229.48 to 269.0	5	Jamkhandi, Bilgi, Sindgi, Mudhol, Indi, Hungund, B. Bagewadi.
Very Low	229.48 and below	1	Muddebihal.

The level of agricultural development in 1978 is slow as there was not a single taluk under Very High agricultural development. Badami and Bijapur were under high developed group. Jamkhandi, Bilgi and Bagalkot under medium; Sindgi, Mudhol, Indi, Hungund and B. Bagewadi fell under low developed group. Overwhelmingly the large part of the district was under low and very low development category. All the eastern taluks of the district including central eastern part were under low and very low development.

1984

Levels of Development	Composite Index Range	No. of Taluks	Name of the Taluks
1	2	3	4
Very High	503.15 and above	1	Badami.
High	439.63 to 503.15	Nil	Nil
Medium	376.11 to 439.63	5	Mudhol Jamkhandi, Bilgi, Bagalkot and Hungund.
Low	3.76.59 to 376.11	3	Sindgi, Muddebihal, and Bijapur.
Very Low	312.59 and above	2	B. Bagewadi and Indi.

The Levels of Agricultural Development (1984)

In the year 1984 Badami taluk is under very high development category. Five taluks (Mudhol, Jamkhandi, Bilgi, Bagalkot and Hungund) are under medium group, while, three taluks (Sindgi, Muddebihal and Bijapur), two taluks (Bagewadi and Indi) are under low and very low development categories. A large number of taluks come under medium development. Yet, about 45 per cent of the area of the district is still under low and very low development group.

The regional disparities in Bijapur district are identified with the help of Kendall's rank order score method considering 28 indicators:

The 28 indicators are grouped as under:

Social Indicators

1. Educational institutions (primary school, high school, and Kinder gartens).
2. All types of Colleges.
3. Post and Telegraphic Offices.

4. Villages with drinking water facilities.
5. Primary Health and Family Planning Centres.
6. Adult Education Centres.
7. Veterinary Hospitals.
8. Film Theatres.
9. Electrified villages as percentage to the total number of villages.

Economic Indicators

1. Number of Commercial Banks.
2. Percentage of Net Sown Area.
3. Percentage of Net Area Irrigated to the Net Sown Area.
4. Percentage of area under HYV to the Net Sown Area.
5. Number of agricultural implements.
6. Number of industries.
7. Length of roads.
8. Length of railways.
9. Number of Agricultural Co-operative Societies.
10. Percentage of Agricultural Workers to the Total Workers.
11. Total number of livestock.
12. Fertiliser consumption.
13. Number of regulated and non-regulated markets.
14. Hydel power consumption.

Demographic Indicators

1. Urban Centres.
2. Density of Population.
3. Percentage of total workers to the total population.
4. Percentage of literacy.
5. Percentage of urban population to the total population.

Methodology

To know the regional disparities and levels of agricultural development in Bijapur district "Combined Rank Score Method" is adopted. The values of each indicator were ranked from 1 to 11 for all the eleven taluks. For calculating combined rank score, Kendall's "ranking co-efficient method" is used. The respective rank order score for 28 selected indicators was posted, added, and then averaged to get averaged rank score. This score was considered as an index of regional disparities. Regional disparities in the levels of development are inversely related to the total rank score i.e., higher the score lower is the taluk on the levels of development, and vice-versa. The ranks of 11 taluks of Bijapur district in respect of individual selected indicators and the total averaged rank order score is worked out and the results found out are as shown in Table 8.1.

Level of Regional Disparities

Classes of regional disparities in the levels of agricultural development (based on combined indicators of social economic and demographic factors).

TABLE 8.1: Levels of development

Classes	Range	Name of Taluks
Very High	Mean - 1 S.D. and more (4.10 and more)	Bijapur
High	Mean to Mean - 1 S.D. (5.24 - 6.38)	Jamkhandi, Hungund, B. Bagewadi, and Badami.
Medium	Mean to Mean + 1 S.D. (5.24 - 6.38)	Sindgi, Mudhol, Muddebihal and Bagalkot.
Low	Mean + 1 S.D. to Mean + 2 S.D. (6.38 - 7.52)	Nil.
Very Low	Mean + 2 S.D. and above i.e., 7.52 and above	Bilgi.

Regional Disparities Based on Combined (28) Factors

Bijapur taluk alone falls under very high development category as it is being a taluka and district headquarter. Jamkhandi, Hungund, Indi, B. Bagewadi and Badami taluks belong to high development regions mainly because of the presence of social factors. The taluks of Sindgi, Mudhol, Muddebihal and Bagalkot get medium development. Among these four taluks Bagalkot and Mudhol are first in the rate of literacy and total number of workers. No taluk comes under low development as the region is progressing economically and socially. But, Bilgi taluka is the only one coming under very low category. It has scored last rank in most of the indicators except the net sown area, agricultural workers and fertiliser consumption.

The relationship between the levels of agricultural development and regional disparities: In order to correlate the levels of agricultural development and levels of regional disparities the Spearman's rank order correlation has been calculated. The '*r*' value has been calculated to the three sectors of regional disparities viz., social sector, economic and demographic sectors with the indices of levels of agricultural development. The '*r*' value of agricultural development with social indicators has been found to be 94.30, with economic indicators 93.55, with demographic indicators 88.64 and with the total indices of regional disparities it is 91.77. This value (91.77) is very high and indicates a maximum significant correlationship. This means that high level of regional development leads to the high development of agriculture and vice-versa. To conclude, the above interrelationship shows that the regional development and agricultural development go hand in hand. Setback in any one of this leads to the setback in another.

An emphasis in the paper is being given on tailoring up of the planning and its implementation strategies to achieve this objective of minimising the regional disparities in the development of agriculture in Bijapur district. Formation of Integrated Area Development Programme to reduce disparities between the taluks and launching over special Integrated Rural Development Programme (IRDP), Special component plans, plans for tribal people and others are some of the latest strategies to be undertaken to develop Bijapur district and its agro-economy.

REFERENCES

Ali Mohammad, 1979 : *Dynamics of Agricultural Development in India*, Concept Publishing Co., New Delhi.

Arogyaswamy, 1971 : "The evaluation and utilisation of Water resources", in Degli V. (ed.), *Natural Resources in the Indian Economy*, Bombay, p. 69.

Bagchi, K., 1979 : "Agriculture in Drought-prone areas" *Geographical Review of India* Vol. 41, No. 4, pp. 293-96.

Bansal, P.C., 1975 : *Agricultural Problems of India*, Vikas Publications, New Delhi.

Bijapur District at Glance, 1984-85.

D.P.A.P. for Bijapur district, 1985-86, Government of Karnataka, Bangalore.

Economic Development of Bijapur District under Drought Prone Areas Programme with World Bank Assistance, Government of Karnataka, 1978.

Jain, B.K.S., 1977 : "Farm Development, New Strategy", *The Economic Times*, July 21.

Joshi, Y.G. and Juliet Dube, 1979 : "Measurement of Regional Disparity of Agricultural Development in M.P." *The Deccan Geographer*, Vol. 22, No. 3, pp. 585-93.

Mohammad, N. (Ed.) : *Perspectives in Agricultural Geography*.

Nadakarni, M.V., 1985 : *Socio-Economic Conditions in Drought-prone Areas*, Concept Publishing Co., New Delhi.

Pal, M.N., 1975 : "Regional Disparities in the levels of development in India" *Indian Journal of Regional Science*, Vol. II, No. 2, pp. 12-35.

9

AN APPRAISAL OF AGRICULTURAL DEVELOPMENT PROGRAMME IN BIJAPUR DISTRICT

G. V. TOGARSI, S. S. NAREGAL AND V. B. KADDI

BIJAPUR DISTRICT of Karnataka State located in the Northern Maidan is well situated in the interior of the Deccan plateau. It is located about 200 miles from the West Coast and is a part of rain-shadow region with erratic and ill-distribution of Southwest monsoon rainfall of 554 mm. of an annual average. As the district belongs to the semi-arid climatic region, it has large variation in the amount and distribution of rainfall, consequently, it is prone to droughts and famines. The district is largest in area (17.069 sq.km.) compared to other twenty districts in the State. It has a population of 24,01,782 (1981) with the density of 141 per km^2. The rate of literacy is 31.96 per cent and the urban population in the district is less, i.e., 24.09 per cent. Out of total population 37.01 per cent are workers and 62.99 are non-workers. Similarly, out of working population the total cultivators are 30 per cent, agricultural labourers 33 per cent, people engaged in household industries contribute hardly 6.03 per cent and other workers 21.16 per cent. The total number of villages are 1,239 and the towns are 19.

The average height of the land is 1,950'. The area is surrounded by semi-arid regions of Karnataka and Maharashtra States. The district has a broad groups of physical divisions with Deccan traps in the north, Kaladgi series in the south-west and peninsular gneissic topography in the east and southeast. The district is drained by five major rivers and is overlain by deep black soils in major portion of the area. The waste land occupies hardly 6.10 per cent (1,04,466 hectares). Climatically the area belongs to semi-arid type with dry and healthy conditions for human settlements. The average annual rainfall of 554 mm. occurs in the region with its large variation in the amount and distribution rendering the district liable to droughts and famines. The mean maximum temperature of the district is 40.00°C and minimum 14.8°C. The district in Karnataka is conspicuous due to its plain and gentle topography and also its geographical entity in the matter of relief, soil, climate, drainage and vegetation. The vegetation cover is very sparse and occupies 4.87 per cent (83,103 hectares) only.

The economy of the district is dwindling due to the occurrence of droughts at least twice in every five years. The present study on the programmes of agricultural development certainly shows the characteristic features of the underdeveloped economy, viz., the growing magnitude of unemployment, poverty, subsistence farming and low standard of living. The land under irrigation is very less, i.e., 10.84 per cent (1,54,258 hectares) compared to the net sown area i.e., 83.09 per cent (14,22,713 hectares). The strength of livestock population is 22,86,734. The total number of land holders in the district is 3,02,566 of which marginal land holders are 9.52 per cent (28,560), small holders 19.60 per cent (59,316), semi-medium 28.97 per cent (87,652), medium 30.86 per cent (93,376) and large holders 11.13 per cent (33,662). About 10.99 per cent of the land was brought under HYV programme in the district. The fertiliser consumption in the district is very low, i.e., 0.013 tonnes per hectare. The number of tractors in the district is 1,807 which emphasises the development of agriculture towards mechanisation. Similarly the high use of electric pumpsets 36,512 suggests the achievement of farmers in the developmental programmes like well irrigation.

The areas which are prone to frequent droughts and scarcities will have these problems in an accentuated manner. Bijapur district is one of the backward districts in the State inspite of the existence of rivers and fertile soils. Though five rivers flow in the district the waters of the river is used only by those villages which are situated on the banks. The Ghataprabha Left Bank Canal irrigates the lands of Jamakhandi (46.52%), Bilgi (56.27%) and Mudhol (70.15%) taluks only. The development of the district mainly depends on the exploitation of underground water and canal irrigation by Almatti Dam (which is not yet completed). Proper development of land and water use thus merit special consideration. The problems concerned to agriculture of small and marginal farmers and agricultural labourers will have to be considered on top priority and the benefits should go mainly to them to strengthen the landuse efficiency. The Jawahar Rozgar Yojana will be quite appropriate to reduce the stress of drought situation as well as unemployment created due to low progress in socio-economic conditions of the study region. The emphasis is also needed to restore ecological balance. It depends upon the development of irrigational facilities, soil and moisture conservation, adoption of bio-technology, afforestation, change in the cropping pattern, agronomic practices, livestock development, rural communications, provision for drinking water supply, development of small and marginal farmers and the agricultural labourers and setting up of agro-based industries. The dry land development programmes, sustainance of animal husbandry, fisheries, horticulture, sericulture and small scale industries (at village level) should accompany the programme.

The main objectives of the programme of agricultural development in Bijapur district is rotating around the drought situation. Therefore, profound plans on permanent basis to promote agricultural development and to generate employment opportunities and to restore ecology are needed.

Agricultural Development Programmes in Bijapur District

Under the agricultural development programme; the Intensive Agriculture Development Programme (IADP) and Intensive Agricultural Area Development Programme (IAAP) are concerned with the promotion of intensive agriculture. This programme was introduced in the district where there are irrigational facilities and the existence of strong input supply agencies. To increase the agricultural production, more and more areas were brought under High Yielding Varieties Programme. Table 9.1 shows this achievement.

TABLE 9.1. Target and achievement under H.Y.V.P.

(Area in Hectares)

Crops	1978		1984	
	Target	Achievement	Target	Achievement
Jowar	27,100	9,674 (35.7 %)	27,500	22,044 (80.16 %)
Maize	35,200	32,290 (91.73 %)	50,600	49,696 (98.21 %)
Wheat	49,000	46,976 (95.87 %)	50,000	37,577 (75.15 %)
Bajra	67,000	14,468 (21.59 %)	70,000	1,49,358 (213.37 %)
Total	1,78,300	1,03,508 (58.05 %)	1,98,100	2,58,675 (130.58 %)

It is observed from Table 9.1 that the percentage of achievement against the target in respect of jowar, maize, wheat and bajra during 1984-85 was 80.16 per cent, 98.21 per cent, 75.15 per cent and 213.37 per cent, respectively. The overall achievement under HYV is 130.58

per cent. The farmers in the district very much depend upon the rainfall for cultivation of different crops due to the uncertainty of rainfall and hardly sufficient irrigational facilities (10.85 per cent of the net sown area).

Soil and Water Conservation

The effects of soil erosion are manifold. The fertility of the soils, the available plant nutrients and organic matter accumulated in the top layer of the soil are eroded away. Consequently the growth of plants and crops deteriorate and agricultural productivity declines. The soil conservation programme in the district is undertaken under DPAP and the target achieved by 1984-85 was 2 lakhs of acres under the programme. Table 9.2 shows this achievement.

TABLE 9.2: Achievements of soil conservation measures

(Area in Hectares)

Year	Target	Bunded Achievement	Target	Surveyed Achievement
1977-78	18,950	24,151 (127.44 %)	30,000	55,495 (184.98 %)
1980-81	8,400	36,203 (430.98 %)	36,400	24,688 (78.82 %)
1984-85	35,000	43,600 (124.57 %)	20,700	19,825 (95.77 %)

The programme will help the unemployed local labourers to get employment. The yield of crops will increase by 50 per cent in the areas of soil conservation and the water table will also rise in those areas. Besides, there will be a reduction in the sedimentation and flood damages, and the moisture content of the soil will increase automatically.

Distribution of Seeds

In order to encourage improved farm techniques farmers have been supplied with better variety of seeds to increase production. For this purpose seed multiplication farm have been set up in the district. Table 9.3 covers the progress made in this regard.

TABLE 9.3: Quantity of seeds distributed

(in quintals)

Crops	1979-80	1980-81	1984-85
Jowar	3,200	1,753	2,204
Wheat	12,600	28,871	37,577

The wheat is grown mostly in irrigated tracts and other assured water supply areas, whereas jowar is grown as dry land crop almost every where in the district.

Technologies

The farmer is capable of and willing to adopt modern techniques, given sufficient motivation yet, new technologies have not been adopted rapidly in all the taluks of the district.

Consumption of Fertilisers

Fertilisers are being used extensively to get better yield. The consumption of fertilisers i.e., Nitrogen (N), Phosphorous (P) and Potassium (K) is shown in Table 9.4.

TABLE 9.4: Consumption of fertilisers

(*In Tonnes*)

Year	N	P	K
1977-78	7,876	2,679	1,873
1980-81	8,116	3,931	2,869
1984-85	8,802	4,415	3,183

The consumption of fertilisers shows steady increase over the period from 1977-78 to 1984-85 mainly because of the modern application of cultivation by farmers.

Manures

Intensification of rural and urban compost, green manure, supply of cuttings and seedlings and their distribution are important activities under the scheme. Table 9.5 highlights the use of various types of manures used in the district.

TABLE 9.5: Use of manures

Year	Urban Compost (in tonnes)	Rural Compost (in tonnes)	Green Manures (in Hect)	Supply of cuttings of seedlings (Nos)	Distribution of green manure (in tonnes)
1977-78	39,921	5,37,654	12,961	2,52,887	215
1980-81	89,791	5,99,643	6,545	2,35,300	163
1984-85	15,442	7,35,797	40,503	10,25,900	201

The use of manures has increased during 1984-85 as compared to

1980-81.

Strategy for the Development of Agriculture

The crux of the problem of arid and semi-arid areas in India to develop agriculture lies in the deficiency of water. The most important solution to it is to utilise the underground water for the use of human and livestock population and also for agriculture. There is a scope for improvement in agriculture through expansion of irrigation facilities along with suitable techniques of cultivation.

Use of Ground Water and Surface Water

It is necessary to make use of the water from tanks and wells for irrigation and a special attention should be given for construction of percolation tanks, check dams and other bunds so as to enable farmers to raise crops and also to enable them to harvest rain water to improve surface water as well as the re-charge of ground water.

Watershed Management

It is another method through which the water supply can be improved in dry land areas. The technology of watershed management consists of :

(i) Storing of rain water in the farm lands in an artificial ponds. This process will reduce the soil erosion and wastage of water. The stored water can be used through pumpsets or through tubes to irrigate the land when rain fails. in Bijapur district this watershed management programme is undertaken in Yarnal village in B. Bagewadi taluk. The programme is guided by the University of Agricultural Science, Dharwad and by ICRISAT. The achievements through watershed management in Yarnal have become very successful.

Exploitation of Ground Water

Ground water has developed rapidly as a source of irrigation through bore wells. Unlike surface water ground water is available to a greater extent in Bijapur district. The utilisation of ground water through tube-wells is increasing day by day to develop agriculture throughout the district. The district has 10,152.66 million cubic metres of potential water. The per capita availability of water in the district is 5.11 thousand cubic metres. The use for drinking and domestic needs is 47.07 million cubic metres. The water used for irrigation is 10,454.50 thousand cubic metres. The water used for industries is 60.62 thousand cubic metres. The annual evaporation is 3,500.6 million cubic metres. The river run off is 3.58 million cubic metres. The potential water resources still unused is 6,590.41 million cubic metres. Therefore, all types of farmers are to be trained and encouraged to practise irrigation at their farm level. The major irrigation project across river Krishna at Almatti and Narayanpur are not completed even after 20 years of their foundation stone. Hence the small scale irrigation schemes at individual farmers level have to be undertaken at massive scale.

Improved Methods of Irrigation

Improved methods of irrigation include sprinkler and drip irrigation. These methods are highly used in dry regions of the world. This method is used in Bijapur district in a limited area that too by rich farmers.

Conservation of Water, Vegetation and Land

A comprehensive and integrated water, soil and forest conservation policy is necessary to improve soil condition, crop yield and the eco-system of the drought-hit areas, such as Bijapur district.

Technology for Dry Land Farming

Evolution of farm technology adopted to the agro-climatic conditions of the dry-hit areas is looked upon as one of the promising strategies. Introduction of suitable high yielding variety seeds, crop combinations, adoption to have mixed farming and livestock farming are the main components of dry land farming technologies. Use of HYV seeds, practice of inter-cropping and mixed farming are to be widely practised on a scientific line in Bijapur district.

The above study of agricultural development programmes in Bijapur district gives sufficient information about the weather phenomena, soil conditions, water resources, land holdings, landuse, cropping pattern, agricultural efficiency and socio-economic features of the farm community, livestock and fodder situations. It is hoped that the indepth study undertaken here on various aspects analysed through several appropriate and latest techniques with maps, tables will certainly help the agricultural planners to improve agriculture economy in Bijapur district. In order to overcome the drought-proneness, the initiating developmental programmes in such regions, one has to observe how far the strategies of irrigation, animal husbandry and rural industry have succeeded and can suggest developmental strategies. In the above context, agriculture in Bijapur district cannot be improved in isolation, rather it can suitably be developed if other dimensions of the region such as social, economic and demographic are taken care of to achieve comprehensive and homogeneous development.

The evaluation of agricultural development programmes in Bijapur district are vast and call for massive planning programmes. The important aspects of the programmes are to make the best use of the available land and water resources and to help for the restoration of ecological balance and agricultural development. The development should be homogeneous and the success of it depends upon the organised and planned development of the area. The benefits of these schemes indicate that the development of the district should be attempted in an integrated manner with time bound programmes.

REFERENCES

Arnon, 1981 : "Modernisation of Agriculture in Developing Countries", John Wiley and Sons, New York, Toronto.

Bansal, P. C., 1975 : "Agricultural Problems of India", Vikas Publications, New Delhi.

Bijapur District Census Hand Book, 1981.

Bijapur District at Glance — 1984-85.

D.P.A.P. For Bijapur District 1985-86, Government of Karnataka.

I.R.D.P. Annual Action Plan 1985-86 (Bijapur District).

Sain, K., 1979 : "Impact of High Yielding Variety on Agricultural Growth", *Commerce*, Vol. 139, No. 3577, pp. 34-36.

Singh, V. R., 1958 : "The Measurement of Land Resources", *Geographical Review*, Vol. 48, pp. 1-9.

10

ALTERNATIVE STRATEGIES FOR AGRICULTURAL DEVELOPMENT IN INDIA

N. K. DE AND A. K. SARKAR

INDIAN ECONOMY is basically dependent on agriculture. Nearly 37 per cent of the GNP of the country is derived from agriculture and allied activities and as large as 61 per cent of our workforce is dependent on it. Agriculture provides livelihood for about 75 per cent of the total population and also accounts for nearly 35 per cent of the total value of country's export. The traditional agricultural practice in India has undergone considerable changes and conspicuous technological innovations have been noticed in this particular sector since mid-sixties of the present century. Although initially the changing phenomena were considered as boon to the national economy as a whole, yet the ultimate result seems to work adversely creating enormous degradation to the total environmental system, both physical as well as social. With this consideration, an attempt has been made in the present paper to review critically, the technological changes which took place in agricultural sector and how far they go adversely and also to prescribe the alternative strategies for optimal agricultural development in India, keeping in view the ecological implications thereto.

Historical Background

Although traditionally agriculture is the most important sector of economic activity in India, yet during the pre-independence period of colonial exploitation, this particular sector did not get due attention for development (Sarkar, 1972). The situation becomes worst when the Indian economy as a whole became shattered during the Second World

War. Moreover, natural growth rate of population went on increasing steadily during those days. In addition to that, millions of mouths had to be filled in, who migrated from both sectors of the then Pakistan. It also changed the landman ratio adversely (Sarkar, 1972). Hence, after independence, in the First Five Year Plan (1951–56) itself, agricultural sector was given due priority. To increase food production projects like Community Development Programme, etc. were initiated in different parts of the country for the first time. In the next phase, by the year 1960, Intensive Agricultural District Programme (IADP) was specially designed to promote a rapid increase in production of food grains in selected districts (Brown, 1971). The programme was assisted by the Ford Foundation, U.S. AID, the Co-operative League of America and government agencies from Japan, West Germany and Denmark. Researchers, who were already engaged in examining the suitability of Indian condition, assured the cultivators to go ahead with high-yielding variety seeds for getting much more production than what indigenous variety could fetch. Thus by the mid-sixties, a new technology evolved in Indian agriculture, which was basically capital intensive in nature (Ram and Trehan, 1981). The new term "Green Revolution" was coined to denote it, although there is controversy about its coverage in terms of spatial dimension as well as its involvement in different types of crops.

Characteristics of Advanced Agricultural Technology

The capital intensive new agricultural technology demands pre-conditional availability of infrastructural facilities in the forms of chemical fertiliser, irrigation water, improved variety of seeds, particularly high yielding variety, plant protection measures, mechanisation in farming, better storage systems, etc. It has been observed that drastic improvement has been recorded during last three decades in all these inputs necessary for sustaining advanced agricultural system.

Fertiliser

Agriculture can be extended: (a) horizontally, i.e. by extending the area under cultivation, (b) vertically, i.e. by increasing the yield rate per unit of land. During the post-independence period, to take more advantage from the later way out, it was decided to apply added doses of fertiliser to soil because it has been estimated that for every tonne of fertiliser nutrient added into the soil, there is an increase of 8 to 10 tonnes of food grains, other things remaining equal. It has also been estimated that about 70 per cent of the growth in agricultural production can be attributed to increase fertiliser use. The use of fertiliser in India has increased so much that now it ranks after USA, USSR and China, fourth in the world in terms of gross fertiliser consumption. From almost zero level per hectare consumption in 1950-51, the consumption in 1986-87 is estimated to have gone up to 84.44 kg. per hectare. The total consumption has gone up from 69 thousand tonnes of nutrients in 1950-51 to about 113 lakh tonnes at present (1989-90 estimate).

But the indiscriminate use of chemical fertilisers has lead to severe consequences. It has been observed that continuous use of chemical fertiliser deteriorates the very soil structure as well as the texture, ultimately making the soil permanently sterile.

Irrigation

More water supply to crop land by means of irrigation system is very much essential for sustaining the advanced agricultural system where added doses of fertiliser is in use. Thus all available sources of water supply viz. canal, river lift, well, tube well (deep as well as shallow), pond and other minor irrigation facilities are in use in considerably greater extent. This is because of the fact that monsoon rainfall, on which the rainfed agricultural practice depends, fluctuates very much, causing natural calamities like two extremes of drought and flood. Moreover, its duration and intensity also fluctuate very much. Hence, attempts have been made to tame major as well as minor rivers so that water reservoirs could be created for supplying water through canal systems at the time of need. Intensive investigations have also been carried out to identify ground water aquifers, so that underground water could also be tapped by means of deep tube well, shallow tube well as well as bore wells.

Improved Variety of Seeds

Government of India recognised the importance of quality of seeds long ago and in the year 1963, the National Seed Corporation was established for meeting the requirements of quality seeds. Distribution of certified seeds in the country has increased manifold during the present decade, from 14 lakh quintals in 1979-80 to about 75 lakh quintals in 1988-89. Apart from good quality indigenous seeds, attempts are being made to introduce more and more the high yielding variety. 'Actually, it is the high yielding dwarf variety of Mexican wheat which triggered off the so-called green revolution of India' (Borlong, 1971). High Yielding Variety of Maize, Jowar, Bajra, Rice, etc. also showed positive results. Area coverage under HYV recorded a steady increase from 19 lakh hectares in 1966-67 to about 550 lakh hectares in 1986-87. The target for the year 1988-89 was about 750 lakh hectares.

Plant Protection

Different types of crops, particularly the HYV, are very much susceptible to insects and pests. That is why it has become a compulsion to protect plants from such misery particularly after introduction of advanced agricultural technology. Attempts are being made for timely detection of pests and diseases through surveillance and judicious use of pesticides in conjunction with other agricultural practices like use of biological, cultural and genetic engineering. Ground and aerial surveillance are regularly undertaken to locate locust breeding grounds by the Central Locust Warning Organisations for launching central operations. The Satellite Laboratory for Remote Sensing and Locust Breeding Grounds, Jodhpur, collects ground data for analysis.

Damages caused by pests and insects are resisted by means of the use of pesticides, which include insecticides, fungicides, weedicide, rodenticide, etc. The humble beginning of manufacturing of pesticides in India commenced with the installation of the first plant in 1952 at Rishra near Calcutta for the manufacture of benzene hezachloride (BHC). At present more than 50 technical grade pescidides are being manufactured in India, whereas more than 100 insecticides have been approved for use. Fortyseven units are engaged in the manufacture of technical grade pesticides and over 500 units are making formulations. Anticipated level of consumption of pesticides for the year 1986-87 was 72,000 tonnes. The demand for various pesticides for the year 1989-90 is estimated to be over 75,000 tonnes.

Farm Implements and Machinery

Mechanised farming is a typical character of the advanced agricultural technology. At present different types of farm implements and machineries are in use in increasing number in Indian agriculture. Tractors, power tillers, harvesters, threshers, etc. are being manufactured in India in greater number, sometimes with foreign collaboration. Here it may be mentioned that increased intensity of cultivation, remunerative agricultural prices, coupled with rising labour costs are a few important factors which persuade the cultivators to increasingly mechanise cultivation. With extension of canal irrigation facilities, sometimes the water table comes very near to the surface and hence, the soil turns out to be weedy and hardy and it becomes difficult to cultivate by means of traditional implements. Due to multiple cropping and HYV programme specially in irrigated areas, the agricultural operations are time bound and quickly spaced which calls for a greater use of agricultural implements.

Storage Facility

Huge amount of agricultural products, produced as a result of implementation of advanced agricultural technology, need to be stored properly, sometimes even as buffer stock. Otherwise, due to lack of storage facility, farmers are to incur losses with their surplus products. To meet the demand for storage of agricultural commodities, government agencies like Food Corporation of India, Central Warehousing Corporation and 16 State Warehousing Corporations have come forward to offer modern storage facilities. Sometimes private agencies also provide storage facilities for agricultural commodities. At present the total storage capacity under management of different government agencies is estimated to be about 125 lakh tonnes.

Effects of Technology on Environment

Advance technology in agriculture was invited in India to increase the production rate to a much higher magnitude. Apparently the result was very much positive in nature. It has been observed that per capita net availability of food grains went up to a level of about 480 grams per day in 1988-89 as compared to that of 395 grams in early fifties.

Government of India has stopped importing major cereals like wheat or rice on government account through Food Corporation of India during 1986. Moreover, the excess production not only helped to create buffer stock but also exporting cereals and other cereal products to other countries. During 1986 export of *basmati* rice continued to be on Open General Licence (OGL-3). In addition to that export of *non-basmati* rice was allowed with effect from 18 February'86 with a limited ceiling. However, no rice was exported by Food Corporation of India. Export of wheat and wheat products (e.g. *maida, suji, ata*, etc.) were allowed within a limited ceiling. In 1986, 5,000 tonnes of wheat was exported to Nepal as commercial sale, 25.62 thousand tonnes to Vietnam as commodity loan and 36.94 thousand tonnes as gift to other countries in addition to 144,80 thousand tonnes supplied to STC and MMTC for export purposes.

Apart from such bright picture in the agricultural sector of India as a whole, some holocaust of dark clouds are also visible at certain corners. Some unwanted and undesirable evils to the total environmental system could be recorded, which need not be allowed to continue.

Firstly, there is no doubt that the use of chemical fertiliser increased the yield rate of various agricultural commodities manyfold, but at the same time it is being observed that continuous use of chemical fertiliser adversely affects physical as well as chemical property of soil. It deteriorates soil structure as well as soil texture, ultimately making the soil permanently sterile, which is of no use for future purposes. Nitrogen is the much needed chemical ingredient to the soil, and 9/10th of it comes from natural sources, whereas only 1/10th is supplied in the form of chemical fertiliser. In a developing country like India, construction of a fertiliser factory is a very costly affair, and is a burden on the national economy. For use of each dose of chemical fertiliser the farmer is to bear extra financial burden as well. Unfortunately, the quantity of the agricultural commodity is being increased with the use of chemical fertiliser, at the expense of quality. It has been observed that the taste of products also goes on deteriorating by application of chemical fertilisers.

Secondly, use of chemical fertiliser in agricultural plot demands more water which is generally supplied by means of different types of irrigation systems. It has been observed that canal irrigation results in occurrence of salt incrustation on the surface of the soil in areas susceptible to salinity. Sometimes the system also results in bringing the water table closer to the surface, which helps the weeds to grow on cropland. Sometimes it also causes problem of water logging (Das and Sarkar, 1972).

Construction of dams and reservoirs involves huge expenditure from public exchequer. Moreover, large areas were submerged under reservoirs and were also destroyed by construction of roads and colonies, that brought untold misery to the poor oustees. Added to these are the problems of salinity and water logging. The net result is the loss of productive potential of land (Khoshoo, 1986). Moreover, some experts are of the opinion that the dams and reservoirs located at different parts of the country may lead to undesirable consequences like enhancement of seismic activities causing flood and other associated hazards like washing away crops, settlements, etc.

Thirdly, to increase yield rate under intensive agricultural programme, high yielding variety of seeds had to be introduced, which are again very much susceptible to insects and pests. To protect plants from these insects and pests, adequate measure in terms of the use of insecticides and pesticides is to be introduced. Unfortunately, whenever such chemical substances are released in the rural environment, there is every possibility of threat to biotic life through terrestial, atmospheric or aquatic pollution. It has been observed that highly toxic chlorinated hydro-carbon elements e.g., DDT, BHC, Endosulfan, Aldrin, Lindane, etc., do not go on changing much if they are applied in the cropland. Moreover, photo-toxicity may be caused from photo-chemical reaction when contaminated leaves and plant tissues are exposed to sunlight. The result is the retarded growth of photo-synthesis process of the plant.

Pesticides not only destroy the pests and insects for which they are applied for, but also pollute the entire surroundings undesirably. As a consequence two-fold hazards are caused: 'firstly, the scope of natural control of harmful insects through biological enemity goes in vain, and secondly, in view of higher resistivity of injurious insects, higher dose of insecticides are required which in turn enhances the danger of pollution' (Basu, 1985). When pesticides are applied at certain place, they do not confine themselves at that particular spot only. On the other hand, in the process of runoff or with irrigation system, they are transported to other areas creating pollution there too. Application of pesticides has also some long run results. The toxic materials contaminate crops, vegetables and grasses. By consuming such vegetables, human beings take some toxic material. Grass and food are consumed by cattle, and in the process contaminated chemicals are accumulated in their body fat as well as in milk. When a person consumes meat or milk of the affected animal, be may develop concentration of toxic materials in the body fat causing diseases like gastro-enterities, allergy, asthma, throat irritation, short sightness, etc. — an overall health hazard.

Fourthly, mechanised farming invites use of agricultural instruments and machineries more and more in one hand, and this capital intensive economy throws away surplus labour on the other. It has been observed that 'in Punjab, one of the pioneering states adopting the new technology, due to big shift towards mechanised power, the use of hired casual labour declined by 44 per cent, and the man days of labour used per hectare of sown area declined from 200 to 120 per annum' (Chowdhuri, et. al., 1969). However, this figure is related to the early eighty's situation, which might have subsequently been aggravated further by this time.

Alternative Strategies Suggested

Severe crisis due to scarcity of foodgrain in India, which was observed during the initial stage of post-independence period, is no more in existence. Now India has become self sufficient in agricultural products, specially food grains. The nation could create sufficient buffer stock of food grains for emergency need. Moreover, at present India is also exporting food items to other countries as stated earlier.

This being the case, now it is necessary to think about alternative strategies to make Indian agricultural more promising and prosperous and to keep it out of necessary evils of advanced agricultural technology, which is being faced by Indian farmers today.

Landscape Ecological Study

It has already been mentioned that Indian economy is based on agriculture, which is being practiced in a rural environment, where this particular economy is being controlled mainly by the physical environment. Cultural environment/management practices, however, is not insignificant in controlling the fertility and productivity of land/soil as such, and hence, agriculture. But it is to be remembered that any cultural/landuse practice must be followed keeping pace with the physical attributes of the land and soil. Cultural (management) practice should not go to such an extent so that the land/soil loose productivity and ultimately become permanently sterile. Here lies the importance of landscape ecological studies, which must be taken into consideration at the very initial stage (Vink, 1974). Thus before prescribing any alternative strategy for agricultural development in India, it is necessary to study the landscape ecology first, on the basis of which the management practices like application of fertiliser, extension of irrigation facilities, use of HYV seed, etc., which have been started earlier, are to be recommended depending upon the physical capability of the land/soil.

From the anthropocentric point of view, the various attributes of the land may be called land resources. The main land resources are climate, parent material, relief, soil, water, vegetation, etc. These various land resources are not at all independent of each other, and their genetic and ecological relations in many respect are wellknown to the ecologists. Their relations differ considerably in different regions. And it is one of the important tasks to study as clearly and as quantitatively as possible the shapes and kinds of these various resources and their relationship within the many different geographically defined tracts of land.

A landscape is a natural arrangement of inter-related tracts of lands, which is very complicated in nature. Accordingly the study of landscape ecology is very much relevant for landuse planning which is very much concerned with the ecology of mankind as well as with that of plants and animals.

The study of landscape ecology sees the land first and foremost as the carrier of eco-system. Eco-system may be natural or cultural ones. In the former, no specific human influence is brought to bear on the processes and phenomena, whereas in the later the process and phenomena occurring within the eco-system have been specifically directed by man to obtain certain results. Landscape ecology describes the various land resources within a given tract of land in so far as they are relevant to their ecology.

The main activity of landuse is that it causes certain cultural eco-system to operate in some manner determined by human intelligence. The landscape ecology of the cultural eco-system is therefore of the utmost importance for agricultural landuse. It studies the way in which various cultural eco-systems (i.e. landuse types) may function within various landscapes, which are sets of living and inanimate attributes, with different kinds of quantities of human controls, i.e. land management. Thus, a landscape ecological study is the study of the attributes of land as objects and variables of eco-systems and of the processes which relates these objects and variables, including a special study of key variable to be controlled by human intelligence.

In this type of studies, terrain, soil and water conditions have a key position with regard to landuse study/planning. The knowledge of land, soil and water condition is indispensable for understanding the past and present landuse as well as for the prediction of future uses. In many regions, these are the primary key variables for human control in cultural eco-system.

Management Aspects of Alternate Strategies

The landscape ecological study of a particular area, as stated earlier, is to be followed by evaluation of resources for optimal use considering both the aspects of social as well as political environment. In the next, phase, planning strategies are to be evolved to manage the physical/social degradation of resources, if any. In the present context, one can suggest the following measures to be taken up to tackle the problems detected so far.

Firstly, instead of using more doses of chemical fertilisers, as an alternate measure one should explore possibilities towards using organic and green manures more and more which otherwise keeps the balance of physical and chemical properties of land and soil intact. Urban garbage, sludge, compost, green manures like *dhaincha*, etc. recorded positive results. In exceptional cases, to supplement nutrients, one can use mixed doses of chemical fertiliser alongwith organic manure, if at all necessary. It has also been observed that land fertility can be enhanced by incorporating farmyard manures or any organic matter like weeds, municipal waste, grass, water hyacinth, cereal straw, finely divided coal, etc. Field experiments show that these organic substances, when mixed with soil or sand or a chemical surface like titania and water undergo slow oxidation in air and releases energy which helps to fix atmospheric nitrogen on those surfaces, thus they become rich in nitrogen. It has also been observed that the method is cheap and easy to work out and can be applied to not only normal soil but also to acidic and alkaline soil condition. The quality of crops produced by this method is better than, that obtained with fertilisers (Dhar, 1972).

Secondly, instead of putting more emphasis on costly multipurpose irrigation system, possibilities should be explored towards less expensive minor irrigation system. It is better not to extend canal irrigation in areas susceptible to incrustation of salt on upper layers of the soil. As an alternative attempt should be made to supply irrigation water from sources like well, tube well, pond, etc. 'Cementing of the pond beds, introduction of tiled drainage, sprinkle irrigation, etc. wherever necessary, should not be ruled out, for minimising loss of water through percolation (Ram and Trehan, 1981). Sometimes it has also been observed that, although the water table reached sufficiently below the normal ground level, yet summer paddy is sown at the expense of drinking water, or water required for other domestic purposes, may be at utmost crisis period. This practice should be restricted. Of course, it does not necessarily mean that one should be against cultivation of summer crop like *boro* paddy. On the other hand, whenever water is available abundantly on the surface, the farmer may be at liberty to use the same for required agricultural practice.

Thirdly, HYV seeds yield more crops, but at the same time they are very much susceptible to insects and pests. To combat the problem, the use of insecticides and pesticides becomes essential at the expense of environmental pollution. It has also been reported that continuous use of HYV seed in a particular area results in compulsory withdrawal of indigenous variety because of the fact that the later variety do not sustain in the changed soil condition. Hence, to bring back the indigenous variety again, attempt should be made to reintroduce the local seeds alongwith organic manure so that nutrients can be added to the soil to withstand the changed condition of the land and soil. This would also keep the environmental condition intact.

Fourthly, the use of implements and machinery makes Indian agricultural activity easy, but at the same time shrinks the job opportunity in the farm land. The pressure of population on land in India is relatively maximum and it is not desirable to adopt capital intensive technology. In rural India the success of land consolidation and co-operative movement could not reach such an extent that compells the farmers to practice mechanised farming. Thus it is pleaded that farm implements and machineries should be used only when they are essential, otherwise agricultural labourers should be allowed to work on the field to earn their bread there from.

Conclusion

By adopting the alternative strategies, the yield per hectare may go down initially, compared with yield rate offered by advanced technology so far, but in the long run, that would definitely yield positive return. Moreover, in terms of cost-benefit analysis, the alternative strategy seems to be more remunerative, considering the net expenditure involved in it. The ultimate consideration of agricultural development in India should no longer be the high output of agricultural products per unit area, but due considerations to be given to safeguard the total environment, both physical and human. This is very much important for saving our future generations to come. For over populous country like India, it is desirable to practice labour intensive cultivation system, rather than capital intensive one. The advanced technology may offer some initial boost to the economy, but ultimately causes enormous environmental degradation, both physical as well as social. Hence, to keep the land and soil permanently productive and to get a steady return out of it for the increasing population, there is necessary to have an ecological approach for landscape study first to be followed by the alternative management aspects as suggested. Lastly, instead of getting excessive agricultural product for the time being and loosing everything in the long run, it is better to opt for the alternative strategies which offer steady output regularly keeping the physical as well as social environment intact.

REFERENCES

Basu, Ranjan, 1985 : Pesticide Residues in Soils of Burdwan District, *Impact of Development on Environment*, The Geographical Society of India, Calcutta, pp. 65-68.

Borlong, Norman E., 1971 : World Cooperation in Green Revolution, *Yojana*, New Delhi, March 21, p. 13.

Brown, Dorris, D., 1971 : Agricultural Development in India's District, Harvard University Press, Cambridge, Massachusettes, p. iii.

Das, B. N. and Sarkar, A. K., 1972 : Rural Area Development, Karnal Area : A Case Study, *Indian Journal of Regional Science*, Vol. IV, No. 2, Kharagpur.

Chowdhuri, T. P. S., et al, 1969 : Resource Use and Productivity on Farms, National Institute of Community Development, Hyderabad, p. 81.

Dhar, N. R., 1972 : World Food Fertility Improvement, The Kamala Lectures for the year 1953, University of Calcutta.

Government of India, 1988 : India 1987 : A Reference Annual, Publication Division, New Delhi.

———, 1988 : The Twenty Point Programme — 1986 : Perspectives and Strategies, Ministry of Programme Implementation, New Delhi.

Khoshoo, T. N., 1986 : Environmental priorities in India and Sustainable Development, Presidential address, 73rd Indian Science Congress Association, New Delhi, pp. 20-21.

Mohan Ram, H. Y. and Trehan, S. K. (Ed.), 1981 : Symposium on Rain Water and Dryland Agriculture, Indian National Science Academy, New Delhi, p. 8.

Sarkar, Asit Kumar, 1972 : Green Revolution in India, *Journal of the Institute of Economic Geography in India*, Calcutta, pp. 49-54.

———, 1980 : Strategy for Optimal Landuse Planning in India : Desired Line of Action, paper presented in Seminar on Strategy for Optimal Landuse in India, Geographical Society of India, Calcutta.

———, 1982 : Landuse Mapping at a Large Scale, *Essays in Thematic Cartography*, Monograph 2, National Atlas & Thematic Mapping Organisation, Govt. of India, Calcutta, pp. 83-85.

Vink, A. P. A., 1974 : Physical Geography for Landuse Planning, Norma Wilkinson Memorial Lecture, Geographical Papers No. 29, Department of Geography, University of Reading, U.K.

11

SPATIAL VARIATION IN LEVELS OF DEVELOPMENT IN U.P.

NOOR MOHAMMAD AND ABDUL MAJEED

INDIA INITIATED the process of planned development right from the inception of First Five Year Plan with a view to bring about a structural transformation of the economy so as to achieve a high and sustained rate of growth, a progressive improvement in the standard of living of the masses leading to eradication of poverty and unemployment and provide the material base for a self-reliant socialist economy. In order to achieve this objective, the emphasis kept on changing during different plan periods depending upon the assessment of overall pace of development. During the first decade of planning, owing to scarcity of resources, and productivity of investment, concentrated efforts were made at those sectors of economy and regions of the country where the returns were likely to be prompt and greatest which brought comparative economic prosperity in aggregate terms at national as well a state levels. But it resulted into lopsided development and regional disparities operationalised through a substantial public sector investment at already developed centres and high incentive rural pocket. In the following decade of planning, therefore, emphasis was laid on balanced regional development. As such efforts were made for development of less developed areas and improvement in the living condition of the society by higher allocation of central assistance to backward areas and promotion of industries.

The strategy of development was also shifted from 'Top Down' to 'Bottom up' and industrialisation of backward areas. District planning was taken up on priority basis and 'area development' and 'target group' approach was launched. In the subsequent decade of planning a new dimension of 'development with social justice' was added up with heavy reliance on 'area development' and 'target group' approach for enhancing the benefits of backward areas and exploited section of the society. 'Command Area Development', 'Drought Prone Area Programme', 'Tribal Area Development', 'Small Farmers Development Agency' are the examples of such programmes. In the Seventh Five Year Plan, the emphasis was laid on improvement of physical quality of life and sustainable development or eco-development. This concept will, perhaps, be carried over in the Eighth Five Year Plan with added emphasis on creation of more employment opportunities for the younger generation.

As a result of 40 years of planned development programme and policies, overall improvement in the economic condition has taken place. The structure of national and state economies has changed. The of technology production in agriculture and industry and the socio-institutional framework within which economic activities are conducted have substantially undergone change. The socio-economic condition of the masses has considerably improved. The literacy level, housing condition, quality of life have gone up. But the level of development has not been uniform at any level. Inter- and intra-sectoral differences in the economic structure became more sharp and noticeable. So is the case with inter- and intra-regional levels of development. As such the regional disparities have kept on aggravating during these plan periods all over the country. And, the State of Uttar Pradesh has not been any exception to this process. There have been overall comparative prosperity among the masses leading to socio-economic transformation specially in rural areas. The structure of the economy has undergone certain changes. But regional disparities have also aggravated here which opens up a vista of research. Therefore, the state of Uttar Pradesh has been taken up for the assessment of level of development.

The study has been conducted with the following aim and objectives:

(i) to identify the levels of agricultural, industrial, social, anthropogenic and economic infrastructural development.

(ii) to assess the spatial variations in the level of development.

The study is essentially based on secondary data collected from different published and unpublished sources. The major sources are:

(i) District-wise indicators of level of development.

(ii) Statistical Abstract of Uttar Pradesh.
(iii) Atlas of School Education.

The levels of development has been assessed for the year 1986 with the only reason that the latest data for all the selected indicators were available only upto this year. The district has been taken as unit of analysis due to: Firstly, a fairly satisfactory data base is available at this level, secondly, it offers better facilities for formulation and implementation of plans as it has a definite administrative set up and fairly well defined geographical boundaries; thirdly, co-ordinated planning and development machinery together with viable training and service facilities are available; and finally, there is better possibility of co-ordination of micro level planning at district level.

For identification, classification, regionalisation and even for exploring casual relationship in explanatory model, as many as 86 indicators, value of which differ from district to district throughout the state, have been selected. These indicators manifest the development which has taken place during last 40 years in respect of different socio-economic parameters. However, five sectors of development have been identified with a number of sub-sectors which are given below:

Major sector		Sub-sector	No.of indicator
A. Agriculture	a.	Agricultural Condition	6
	b.	Agricultural Infrastructure-I	5
	c.	Agricultural Infrastructure-II	6
	d.	Agricultural Production	4
B. Industry	a.	Industrial Condition	4
	b.	Industrial Production	3
C. Socio-institutional	a.	Educational Institutions and Services-I	4
	b.	Educational Institutions and Services-II	4
	c.	Educational Institutions and Services-III	4
	d.	Medical Institutions and Services-I	3
	e.	Medical Institutions and Services-II	3
	f.	Medical Institutions and Services-III	6

Contd.

D. Demographic	a.	Urbanisation	4
	b.	Literacy Level	2
	c.	Enrolment ratio in schools	8
	d.	Enrolment of Girls in schools	3
E. Economic Infrastructure	a.	Transport	4
	b.	Communication	6
	c.	Power and Electricity	4
	d.	Housing conditions	5

In selection of large number of indicators mainly two basic problems arise viz. problem of collinearity and conversion of these indicators into comparable unit and contracting them into few indices in order to compare inter-district development levels. For the former correlation matrices have been worked out and the indicators having very high level of correlation have been deleted. The latter has been a matter of research for various disciplines for the last three decades. However, the major methods developed for this purpose are preparation of composite index with the help of scores, ranking methods and principal component analysis/factor analysis. All these methods have their own merits and limitations. A large number of studies have been done to work out composite index of development either by simple aggregation of ranking of areal units, or by assigning weights to selected indicators.

In the present study a combination of both is followed. Firstly, all the indicators have been converted into comparable unit by standardising them with the help of following formula which is also known as *Zi* value or '*Z*' score:

$$Z_{ij} = \frac{x_{ij} - \bar{x}_i}{\delta x_j}$$

where:

Z_{ij} = standardised value of the indicator,
x_{ij} = value of x_j variable in this '*i*'th observation,
$\bar{x}_i$ = means value of x variable
δx_j = standard deviation of x_j variable.

After working out the '*Z*' score of all the indicators, average is taken out for each group of variables in all the five vectors. Similarly, average values have been calculated for all the five vectors which give the level of development of each sector. Since all the five sectors are not equally significant nor they equally contribute to the socio-economic structure of the state, weightage has been given to each of them as noted below:

Agriculture	0.35
Industry	0.25
Economic Infrastructure	0.20
Socio-institutional	0.10
Demographic	0.10
Total	1.00

In order to assess over all level of development, the sector averages have been multiplied with its respective given weight. The resulting figures of all these averages have been added together and divided by five which gives a composite index of development for each district.

The Geographical Profile of the State

The land locked state of Uttar Pradesh, extending over 23° 52′ to 31° 28′ N. Lat. and 77° 04′ to 84° 38′ E. Long. covers an area of 294, 363 Km^2. Structurally it constitutes the part of Indo-Ganges plain which lies between Peninsular India in the south and recently built Himalayan mountain in the north. The plain belongs to the recent period of the geological history and is formed by the sedimentary deposits brought by the rivers of the Himalayas in the north and the Peninsular India in the south (Krishnan, 1960). The materials brought by these rivers ranges from big boulders to fine silt and clay. Geologically it may be classified into two divisions: the older alluvium called bhangar, and newer alluvium known as *Khadar*. The former is found at higher elevation and is free from flood, while the latter stretches along the river basins and are occasionally flooded during the wet monsoon months. The entire state is mainly drained by the rivers of Ganga, Yamuna, Gomati, Ghaghara and Rapti which play a significant role in agricultural operations (Fig. 11.1).

On the basis of structure and relief the state is divided into four physical divisions viz. The Montane Tract, The sub-Montane Tract, The Ganga Plain, and the Trans-Yamuna Tract. The Montane tract lies in the extreme northern part of the state including all the hill districts,

FIG 11.1: Location of study area.

while the sub-montane tract consisting of two narrow belts of *Bhabhar* and Tarai. stretches along the foothills of Himalayas from the district of Bijnor to Gorakhpur. The Ganga plain covering substantial part of the state occupies the central portion and is most fertile tract, while the trans-Yamuna tract known as Bundelkhand region, lies in the south of river Yamuna and has rocky and infertile soil interpersed with patches of black soil.

The state enjoys sub-humid climate with two monsoon seasons: southwest monsoon season, and northeast monsoon season. The former extending from mid-June to October is influenced by the humid winds of the oceanic origin and is characterised by the cloudy weather, high relative humidity and heavy rainfall, the latter is distinguished by dry winds of continental origin, extremes of temperature, clear sky and low humidity. The two monsoons closely correspond with kharif and rabi seasons respectively. The state has variety of soils but more than two-thirds of it is fertile alluvial soil and almost one-third is hill and Bundelkhand soil which are infertile.

It is essentially an agricultural state where more than 80 per cent population depends on agriculture directly or indirectly. Although there are other sources of economy, agriculture is its backbone. The state lacks in power and mineral resources, and hence, in industry as well. Therefore, the economic development and prosperity of the masses mainly depend on sound agricultural base. About two-thirds of the total area of the state is under cultivation and most of it is irrigated. Kharif and rabi are two main agricultural seasons. Rice, maize, jowar and bajra are the major crops of the former, while wheat, peas, gram, barley and potato are the principal crops of the latter.

Uttar Pradesh is one of the most densely populated state of the country supporting a population of eleven crores with an average density of 377 persons per Km^2. It has a very complex population structure specially in case of caste composition. More than hundred castes are found here but the major ones are Brahmin, Rajput, Shaikh, Jat, Kurmi, Scheduled castes, Kori and Kachhi. The state with an average literacy index of 27.4 per cent ranks lowest among Indian states. The literacy per cent among males is higher (38.9) than females (14.48). It has a fairly good network of means of transport and communication, although there are a lot of regional variations.

Spatial Patterns of Level of Agricultural Development

In spite of multiferous development in secondary and tertiary sector of economy, primary sector still contributes substantially to the national income and the lion's share of it comes from agriculture which has recorded considerable development during last three decades due to massive induction of new farm technology. In the present study agricultural development has been considered to be the function of 21 indicators which have been grouped into four categories. The composite index of agricultural development based on the aggregation of these four categories is given in (Table 11.1).

TABLE 11.1: Levels of agricultural development

Category	Score Range	No. of districts	% age to total districts
High	> 0.78	10	17.86
Medium	0.78—0.40	19	33.93
Low	0.40—0.00	17	30.35
Very low	0.00—0.40	10	17.86
Total		56	100.00

Figure 11.2 shows the spatial distribution of region of different levels of development. With the exception of Fatehpur district, the region of high development form a contiguous belt covering the districts of Saharanpur, Muzaffarnagar, Meerut, Ghaziabad, Bulandshahr, Bijnor, Nainital, Rampur and Pilibhit. A strict scrutiny of different selected indicators shows that these districts are, by and large, very favourably placed with regard to these indicators. The group averages of the four categories also point towards their condusiveness. The district of Fatehpur has come to high level of development due to the high value of some of the indicators. The region of moderate development, in fact, gets projected towards south from the core of high development region with the exception of one region located in the extreme eastern Uttar Pradesh. More than one-third district fall under this category and they form mainly two regions with the exception of Dehra Dun which has also moderate level of agricultural development. The major region covers 11 districts while there are only 4 districts in other region. In the regions of moderate level of agricultural development, the indicators of agricultural conditions and infrastructure were having higher values. With the exception of Lalitpur district the region of low agricultural development form a contiguous belt covering 16 districts of northeastern Uttar Pradesh.

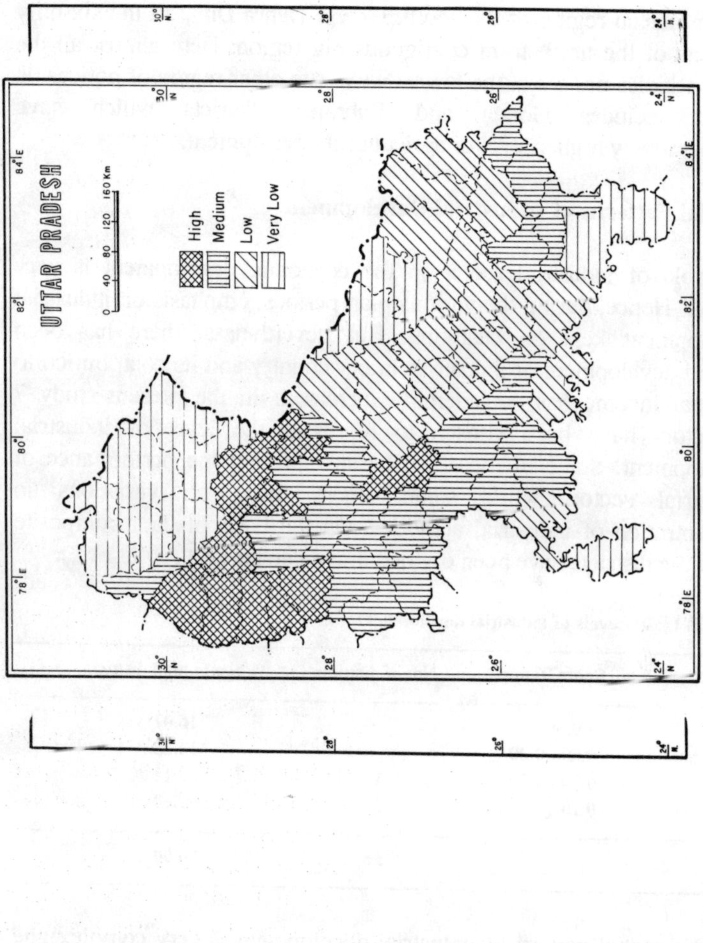

FIG 11 2 Levels of agricultural development.

Very few indicators rank high in these districts. Most of them have moderate to very low status. The region of very low agricultural development is either confined to hill districts of the north with the exception of Bahraich and Gonda districts of trans-Ghaghara plain, or to the plateau region of the south. Except Dehra Dun, all the six hilly districts of the north form contiguous big region. Here almost all the chosen indicators are at the lowest web. The other region of noticeable extent includes Gonda and Bahraich districts which have comparatively higher level of agricultural development.

Spatial Patterns of Industrial Development

The role of industrial sector in the economic development is very crucial. Hence, during the various plan periods, emphasis on industrial development kept on changing. But nevertheless, there has been overall development of industries in our country and its contribution to national income has considerably increased. In the present study 7 indicators have been chosen to represent the level of industrial development. Some of these indicators represent the performance of industrial sector, while some others focus on variations in concentration of industrial activities. On the basis of their composite index the districts have been divided into 4 categories (Table 11.2).

TABLE 11.2: Levels of industrial development

Category	Score Range	No. of districts	% age to total districts
High	> 0.60	9	16.07
Medium	0.60—0.20	8	14.28
Low	0.20—0.20	18	32.15
Very Low	—0.20	21	37.50
Total		56	100.00

The spatial pattern of industrial development is very complex and uneven (Fig. 11.3). On the whole the state is at low web and majority of districts have very poor industrial base. There are only 9 districts scattered in 9 patches over the state having high level of industrial development. A cluster of 4 districts is located in western U.P. Here 5 out of 7 indicators rank high. The other region including the districts of Mirzapur, and Allahabad rank high in majority of selected indicators.

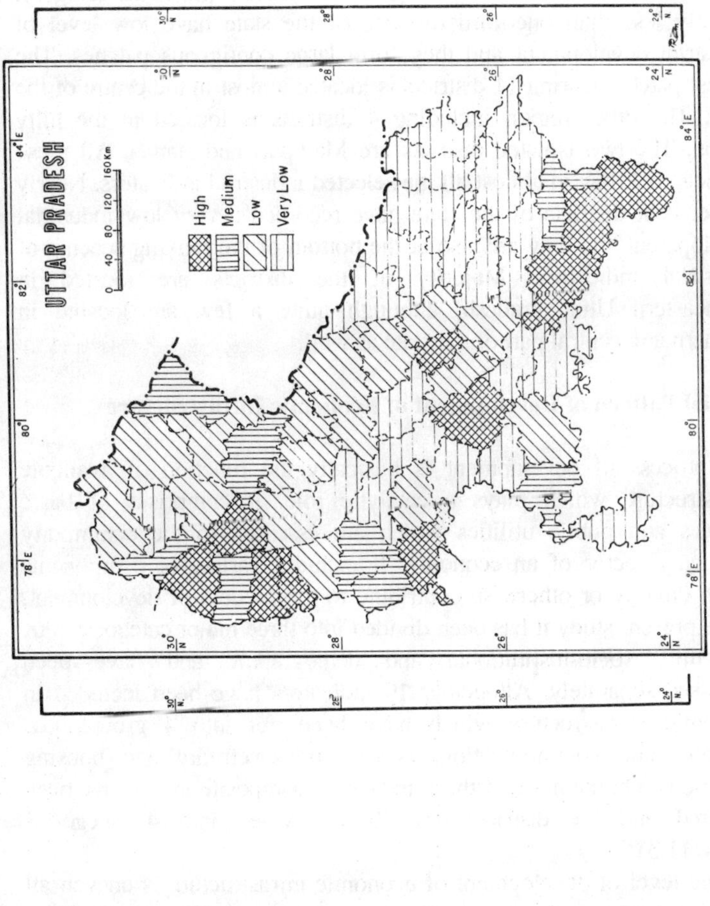

FIG. 11.3: Levels of industrial development.

Besides these, Agra, Kanpur and Lucknow are the other districts having high industrial development. The districts with moderate industrial development are even more widely scattered. Except a single patch of two districts (Aligarh and Mathura), the other regions consist of a single district. There are altogether 8 districts under this category. Slightly less than one-third districts of the state have low level of industrial development and they form large contiguous patches. The largest patch covering 11 districts is located almost in the centre of the state. The other region including 4 districts is located in the hilly region. The two isolated districts are Mainpuri and Banda. All these districts rank low in almost all the selected industrial indicators. Nearly 40 per cent districts of the state have recorded a very low industrial development and they are also at the bottom of the ranking scheme of industrial indicators. Majority of the districts are located in northeastern Uttar Pradesh, although quite a few are located in southern and central part of the state as well.

Spatial Pattern of Development of Economic Infrastructure

The process of development is basically the function of available infrastructure which plays a catalytic role. It comprises of basic services and public utilities which are essential to the commodity producing sector of an economy. It includes variety of components which directly or otherwise contribute to the process of development. In the present study it has been divided into three major categories viz. economic, socio-institutional and demographic and have been discussed separately. Altogether 19 indicators have been included in economic infrastructure which have been put into 4 groups viz. transport and communication, power and electricity and housing conditions. On the basis of these indicators composite index have been prepared and the districts have been divided into 4 categories (Table 11.3).

The level of development of economic infrastructure is uneven all over the districts and present a very complex picture (Fig. 11.4). There are 9 districts with high level of development and are found in four pockets. The largest region is located in western U.P. covering 5 districts which rank high in almost all the selected indicators. The other region comprising the district of Lucknow and Rai Barely is not very comfortable. Bijnor and Agra are the other districts of this category and their position is also somewhat peripheral with regard to the

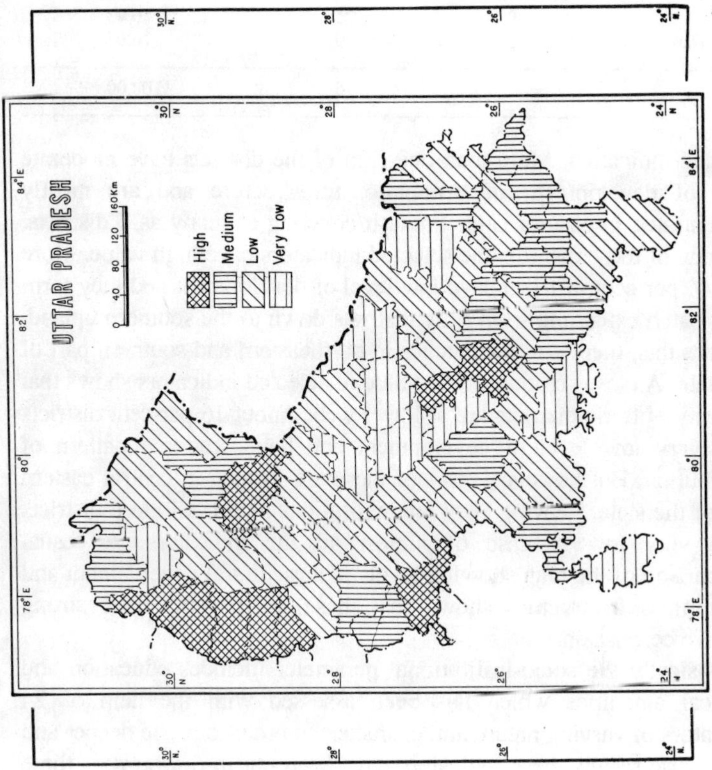

FIG. 11.4: Levels of development of economic infrastructure.

TABLE 11.3: Levels of development of economic infrastructure

Category	Score Range	No. of districts	% age to total districts
High	> 0.70	9	16.07
Medium	0.70—0.30	12	21.43
Low	0.30—0.10	26	46.43
Very Low	< – 0.10	9	16.07
Total		56	100.00

selected indicators. More than one-fifth of the districts have moderate level of development of economic infrastructure and are mostly concentrated in eastern Uttar Pradesh covering as many as 7 districts. In many of these districts the selected indicators have high value. More than 46 per cent districts have low level of development and they form a big patch extending from hilly districts down to the southern upland. Besides this, there are few patches in northeastern and southern part of the state. A close scrutiny of the value of selected indicators shows that majority of these districts are at lower web. About 16 per cent districts have very low level of development and have scattered pattern of distribution. But however, most of these are located in central eastern part of the state. Leaving very few exceptions a part, all these districts stand very low in case of almost all the indicators. A casual comparison of the map showing levels of agricultural development and economic infrastructure shows that these components have strong positive correlation.

Basically the socio-institutional parameter includes education and medical amenities which has been assessed with the help of 24 indicators of varying nature and character. In order to probe deeper and get precise results these indicators have been put in 6 groups — three each for educational institution and services and health institution and services. A composite index for each district has been worked out on the basis of these indicators and the districts have been divided into 4 categories (Table 11.4).

TABLE 11.4: Levels of socio-institutional development

Category	Score Range	No. of districts	% age to total districts
High	> 0.60	9	16.07
Medium	0.60—0.20	11	19.64
Low	0.20—0.20	21	37.50
Very Low	< —0.20	15	26.79
Total		56	100.00

The level of socio-institutional development represents unexpected pattern (Fig. 11.5). Out of 9 districts of high level of development 3 are located in a contiguous belt in the hilly region which is a bit surprising. The other six districts are located on the peripheries in patches starting from western to eastern U.P. However, all these districts have highest value in case of majority of selected indicators. The region of moderate development covering around one-fifth of total districts, is highly concentrated in south western part of the state. The other regions are scattered in 5 pockets covering one district each. The regions of low level of development covering the largest number of districts are found in central and northern part of the state. These districts, by and large, rank low in cases of selected indicators. The region of very low level of development covering more than one-fourth of the districts forms a big contiguous chunk comprising of 12 districts which are located in eastern U.P. The remaining 3 districts are Budaun, Shahjahanpur and Bijnor which is somewhat surprising as these districts are, by and large, at moderate level of development in several respects.

Levels of demographic development constitute one of the important segments of the overall development of a region. In fact the entire process of development is turned towards the improvement of the quality of life. In the present study 17 indicators are selected which have been put into 4 groups viz. urbanisation, literary level, enrolment ratio in schools and school education of girls. On the basis of these indicators composite index has been prepared and the districts have been divided into 4 categories (Table 11.5).

The spatial pattern of distribution of levels of demographic development represents unexpected and surprising results (Fig. 11.6). More than 20 districts have high level of development. With the exception of Uttar Kashi and Tehri Garhwal, all the districts of the state have high level of demographic development which has mainly resulted due to high enrolment ratio and literacy level. The other region of high development covers the districts of Farrukhabad, Kanpur and Etawah which rank very high in almost all the selected indicators. Besides these, there are 3 more districts—Meerut, Lucknow and Ghaziabad with high level of development. In these districts urbanisation index, literary level and enrolment ratio is very high. The regions of moderate level of development cover one-fourth districts of the state but have very uneven distribution. These are found in south and western peripheral areas of the state and the largest chunk is in western U.P. covering 7 districts. Eastern Uttar Pradesh is the other

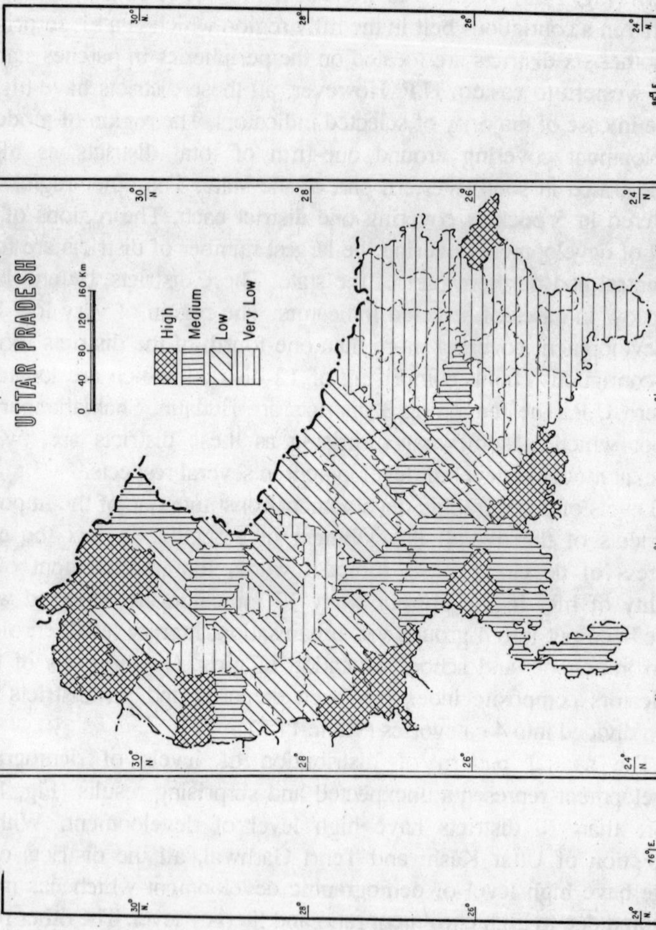

FIG. 11.5: Levels of socio-institutional development.

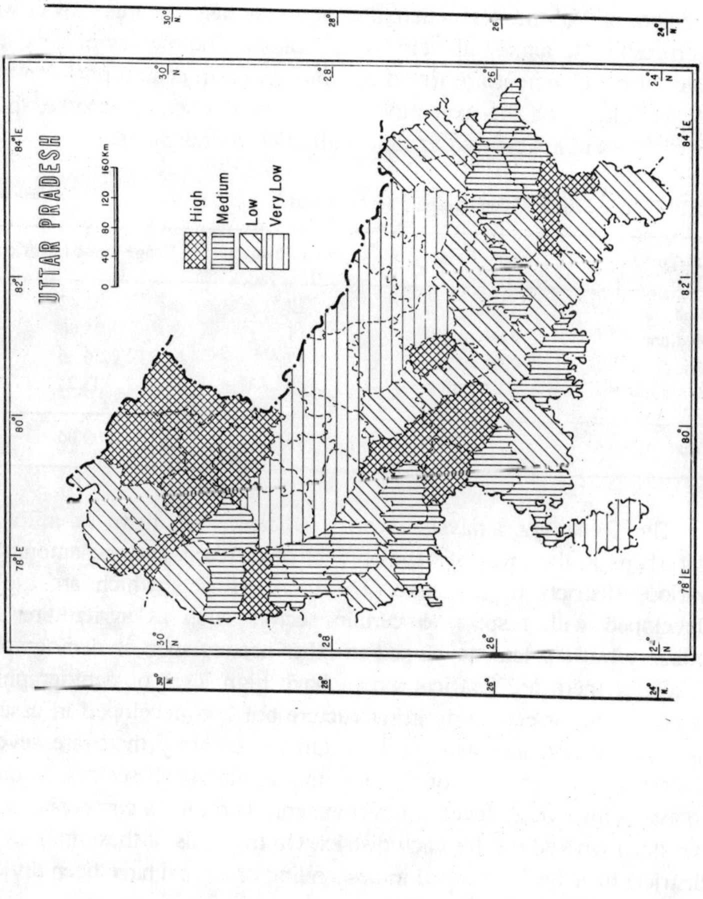

FIG. 11.6: Levels of demographic development

region covering 3 districts of Ballia, Ghazipur and Azamgarh. All these districts have moderate to high value of all the selected indicators. The region of low level of development covers about 30 per cent of the districts and are mostly located in southern and northwestern part of the state. The largest region comprising of 10 districts is found in southern part of the state where almost all the districts are at lower web with respect to almost all selected indicators. The regions of very low level of development are found in a big chunk in central northern part of the state, with the exception of Lalitpur district. Of course, these districts are at lowest web in almost all selected indicators.

TABLE 11.5: Levels of demographic development

Category	Score Range	No. of districts	% age to total districts
High	> 0.60	12	21.43
Medium	0.60—0.20	14	25.00
Low	0.20—0.20	17	30.36
Very Low	< —0.20	13	23.21
Total		56	100.00

The foregoing analysis clearly indicates that there is a lot of variations in the level of development of different sectors. among the various districts of the state. There are districts which are highly developed with respect to certain sectors such as agriculture and industry but are less developed in other sectors such as demography Likewise there are districts which have high level of demographic—development or economic infrastructure but less developed in case of agricultural and industrial sectors. On the contrary there are several districts with high level of development in almost all sectors. In order to assess the overall level of development, therefore, a composite index has been worked out for each district. On the basis of these indices the districts have been arranged in descending order and have been divided into 4 categories (Table 11.6).

The level of development, in fact, is the aggregate picture of the attainment of the district with respect to various selected sectors. The spatial pattern of development is uneven and very complex (Fig. 11.7). More than one-fifth districts—scattered in many pockets, have attained high level of development. There are six pockets consisting of one district each where the level of development is high. These are Agra

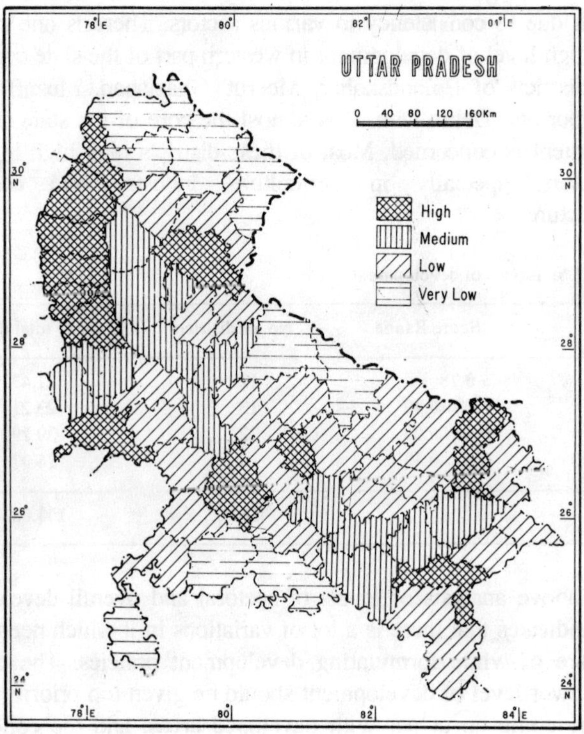

FIG. 11.7: Levels of development.

Kanpur, Lucknow, Varanasi, Bijnor and Gorakhpur. The first 3 districts owe their high level of development to industrial and demographic development, while Bijnor attains it due to economic infrastructure, demographic and agricultural development. It is due to industrial and agricultural development in Varanasi, while Gorakhpur attains it due to consistency in various sectors. There is one compact belt of high level of development in western part of the state consisting of the districts of Bulandshahar, Meerut, Ghaziabad, Muzaffarnagar, Saharanpur and Dehra Dun. It is almost the core of the state as far as development is concerned. Most of these districts rank high in almost all sectors, specially in agriculture, industry and economic infrastructure.

TABLE 11.6: Levels of development

Category	Score Range	No. of districts	% age to total districts
High	> 0.78	12	21.43
Medium	0.78—0.40	13	23.21
Low	0.40—0.00	22	39.29
Very Low	0.00—0.40	9	16.07
Total		56	100.00

The above analysis of levels of sectoral and overall development clearly indicates that there is a lot of variations in it which needs to be taken care of while formulating development policies. The districts having lower level of development should be given top priority so that they may come up at par with developed areas, and the concept of planning with social justice may be fulfilled.

REFERENCES

Dhar, P. N. and Sastry, D. K., 1969 : *Inter-State Variations in Industry in India* 1951-61. Institute of Economic Growth, *Mimeogrophy*.

Dube, R. S. and Misra, R. P., 1981 : Levels of Education : A Versatile Indicator of Regional Development, *Geographical Review of India*, Vol. 43.

Gosal, G. S., and Gopal, K., 1984 : *Regional Disparities in Levels of Socio-economic Development in Punjab*, Kurukshetra.

Gulati, S. C., 1977 : Dimensions of Inter-district Disparities, *Indian Journal of Regional Science*, Vol. 9, No. 2.

Kulkarni, K. M., 1977 : Micro Level Variations in Economic Development in Rajasthan, *Indian Journal of Regional Science*, No. 2, Vol. 9.

Misra, S. N. and Chopra, A., 1973 : Dimensions of Inter-district Disparities in Levels of Development in M.P. — 1971, *Indian Geographical Journal*, No. 1, Vol. 53.

Mitra, A., 1964 : *Levels of Regional Development in India*, Census Publication, Delhi.

Nath, V., 1970 : Levels of Economic Development and Rates of Economic Growth in India : A Regional Analysis, *National Geographical Journal of India*, Vol. 16.

Pal, M. N., 1962 : *Composite Index of Economic Development : A Method of Regional Analysis of Economic Development with Special Reference to South India*, Regional Survey Unit, Indian Statistical Institute, New Delhi.

Patnaik, S. and Chattopadhya, R. N., 1975 : Spatial Variations in Levels of Development in Orissa, *Indian Journal of Regional Science*, No. 1, Vol. 7.

Prakash, S., 1977 : Regional Inequalities and Economic Development with Special Reference to Infrastructural Facilities in India. *Indian Journal of Regional Sciences*, No. 2, Vol. 9.

Rao, A. V. V. S. K., 1980 : 'Inter-state Disparities in Industrial Development' in H.S. Yadav (Ed.). *Planning, Regional Development and Disparities in India*, Criteria Pub., New Delhi.

Rao, S. K., 1973 : A Note on Measuring Economic Distance between Regions in India, *Economic and Political Weekly*, April 28.

Raza, M., 1978 : 'Levels of Regional Development in India', Paper presented at Indo-Soviet Symposium on Regional Development and National Planning.

Sampath, R. K., 1977 : Inter-State Inequalities in Income in India : 1951-71, *Indian Journal of Regional Science*, No. 1, Vol. 9.

Sastry, D. U., 1970 : Inter-State Variation in Industry in India, 1951-61: A Tentative Explanation, *Indian Journal of Regional Science*, No. 1, Vol. 2.

Sengupta, P. and Sadaryule, G., 1968 : 'Economic Regionalisation of India : Problems and Propects', *National Geographical Journal of India*, Vol. 16.

Sharma, K. L., 1975 : 'Spatial Disparities in Rajasthan', *Indian Journal of Regional Science*, No. 1, Vol. 7.

Sharma, P. C., 1981 : Patterns of Industrial Development in M.P., *Indian Journal of Regional Science*, No. 1, Vol. 13.

Srivastava, S. L., 1983 : Regional Disparities in Agricultural Development in M.P., *Indian Journal of Regional Science*, No. 2, Vol. 15.

Tewari, P. T., 1984 : *Changing Pattern of Development in India*, Delhi.

Yadav, J. S. and Prasad, H., 1966 : Spatial Pattern of Economic Development in India, *Deccan Geographer*, No. 2, Vol. 4.

12

LEVELS OF AGRICULTURAL DEVELOPMENT AND ITS CORRELATES

V. S. BHADRAPUR AND S. S. NAREGAL

AGRICULTURAL DEVELOPMENT in a true sense, denotes the quality of the agricultural system of a region in terms of productivity, diversification and commercialisation consistent with a desired state of agrarian relations and ecological balance. The level and the rate of agricultural development may also be distinguished: the former represents a picture prevailing at a particular point in time, while the latter stands for the progress achieved over a given period. If the process of agricultural development is regulated in a systematic line, it becomes agricultural planning. The development of agriculture is to be judged from the degree of equity in farm incomes and the nature of agrarian relations. The test of any development is the extent to which it delivers social justice

Although there is a lot of agricultural transformation on planned lines the precise definition of agricultural development is not yet available. This is because agricultural development is a much more comprehensive concept than is normally understood. In fact, much research work has been done on the development of agriculture by geographers but the essential meaning of the term is nowhere given as a many of them are confined only to agricultural productivity. Of course, productivity is one of the vital aspects of agricultural development. Therefore, in this context scholars have tried to understand the patterns and processes of the crucial problem of agricultural development as it is a multidimensional concept.

Study Area

Bellary district covers an area of 9,898 km² and is located in the central portion of the eastern sector of the Karnataka state. It lies between 14° 35′ and 15° 50′ N. latitudes and 75° 40′ and 77° 11′ E. longitudes. The district is bounded on the north by Raichur district, on the west by Dharwad district and on the south and south-east Chitradurga districts of Karnataka State, while on the east Anantapur and Karnool districts of Andhra Pradesh are located (Fig. 12.1). The district has 589 villages with a population of 1.5 million (1981 Census). The district has eight talukas (Bellary, Hadagalli, Hagaribommanahalli, Harpanahalli, Hospet, Kudligi, Sandur and Sirguppa) and two revenue sub-divisions (Bellary and Hospet). The district had altogether 6,10,168 workers out of which 4,42,746 were agricultural workers (cultivators + agricultural labourers). Agricultural workers alone account for 29.73 per cent to the total population and 72.56 per cent to the total workers. Red loamy and black are the two main soil groups of the district. Climatically the district is situated in an arid zone of India. It has a scanty rainfall of about 575 mm. per annum which precipitates in 39 days and these rainy days are distributed over a period of six months. December is the coolest month when a mean maximum daily temperature is 29.7°C and a mean minimum daily temp. is 16.7°C. On the other hand, April happens to be the hottest month when the mean maximum daily temp. is 39.2°C and the mean minimum daily temp. 25.2°C. The district is noted for its hot summer and dry weather for the major part of the year. The river Tungabhadra is the backbone of the agricultural economy of the district which borders both in the west and in the north. Near Hospet across this river a dam was built in 1954 to irrigate the dry lands of the district. But it can irrigate only three talukas viz., Hospet, Bellary and Sirguppa* and the remaining talukas have to depend upon rainfall and other sources of irrigation for cultivation of the crops. Droughts are the common phenomena.*

Out of 9,898 km² of the geographical area of the district about 61.51 per cent is under cultivation. During the study period in the year 1975-76 net sown area was 6,06,270 hectares (63.25%) whereas the net sown area increased to 6,11,275 hectares (64.02%) to the total geographical area in the year 1985-86. This might be due to reclamation of wasteland for cultivation.

* Most of the stored water from Tungabhadra dam goes to Andhra Pradesh, due to the slope of the topography and hence it is a joint project of Govt. of Karnataka and Govt. of A.P.

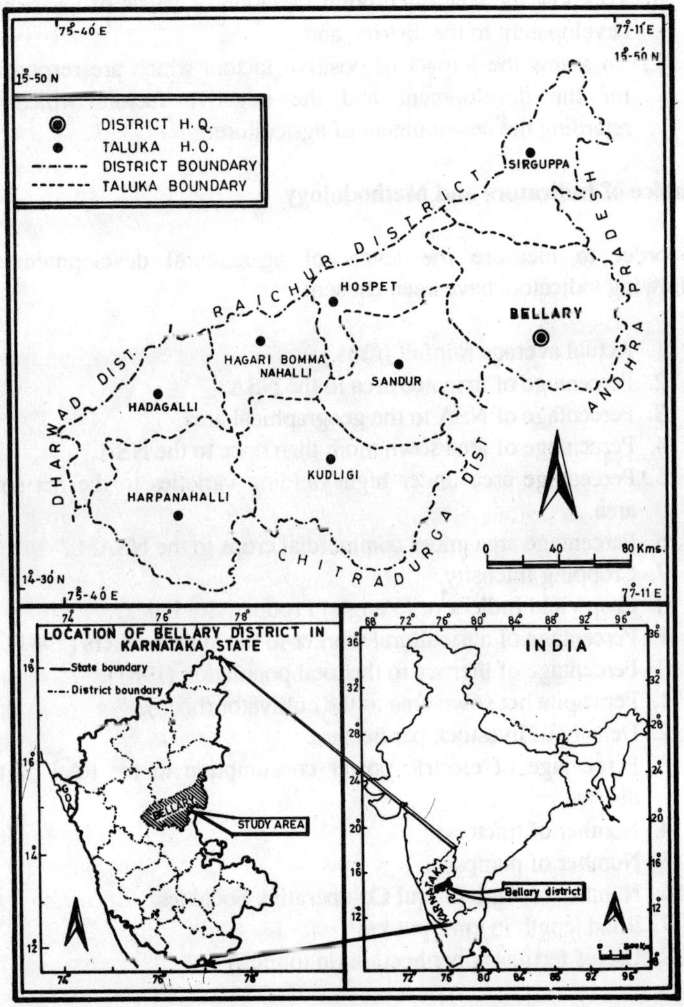

FIG. 12.1: Administrative divisions: Bellary district.

Aim and Objectives

The main objectives of the study are as follows:

1. To know the spatio-temporal variation of levels of agricultural development in the district, and
2. To assess the impact of positive factors which are responsible for the development and the negative factors which are retarding the development of agriculture.

Choice of Indicators and Methodology

In order to measure the levels of agricultural development the following indicators have been chosen:

1. Actual average rainfall (Cms.).
2. Percentage of irrigated area to the NSA.
3. Percentage of NSA to the geographical area.
4. Percentage of area sown more than once to the NSA.
5. Percentage area under high yielding varieties to the net sown area.
6. Percentage area under commercial crops to the NSA.
7. Cropping Intensity.
8. Crop yield Indices or (Yang's) Productivity Index.
9. Percentage of agricultural worker to the main workers (1981).
10. Percentage of literacy to the total population (1981).
11. Per capita net sown area to the cultivator (hect.).
12. Density of livestock per hectare.
13. Percentage of electric power consumption to the total of the district.
14. Number of tractors.
15. Number of pumpsets.
16. Number of Agricultural Co-operative Societies.
17. Road length in kms. per km^2.
18. Use of fertilisers per hectare (in tonnes.)

In fact there are a number of methods to measure the levels of agricultural development but they suffer from one or the other shortcomings.

In view of these shortcomings, the Kendall's Rank Order Score Method is adopted. This method is very simple in calculation but is the most appropriate in its results. Accordingly the ranks for each indicator are awarded vertically and summed up horizontally (talukawise). The total rank score of each taluka is divided by eighteen which gives an average score. Thus the results obtained for both the study periods are classed with the help of mean and S.D. method which gives five groups of developmental regions. Further in order to know the changes of levels of agricultural development the results of both the years are compared. These results are again arbitrarily grouped to know the variation. In order to know the relationship between levels of agricultural development and the individually eighteen factors the Spearman rank order correlation is worked out.

Spatial Pattern of Levels of Agricultural Development

During the year 1975-76 there was not a single taluka under the category of very high development of agriculture, whereas under the high category there was only Hospet taluka (Table 12.1). Bellary, Sirguppa and Hadagalli were under medium category of development. The two talukas viz. Kudligi and Harpanahalli were under low category of agriculture development.

TABLE 12.1: Taluka-wise total rank and average rank score

Taluka	1975-76		1985-86	
	Total Rank	Average Rank Score	Total Ranks	Average Rank Score
Bellary	69	3.83	78	4.34
Hadagalli	71	3.94	74	4.11
H.B. Halli*	96	5.33	88	4.89
Harpanahalli	87	4.83	83	4.61
Hospet	53	2.94	56	3.11
Kudligi	81	4.50	97	5.36
Sandur	105	5.83	92	5.06
Sirguppa	70	3.89	69	3.83

* Hagribommonahalli.

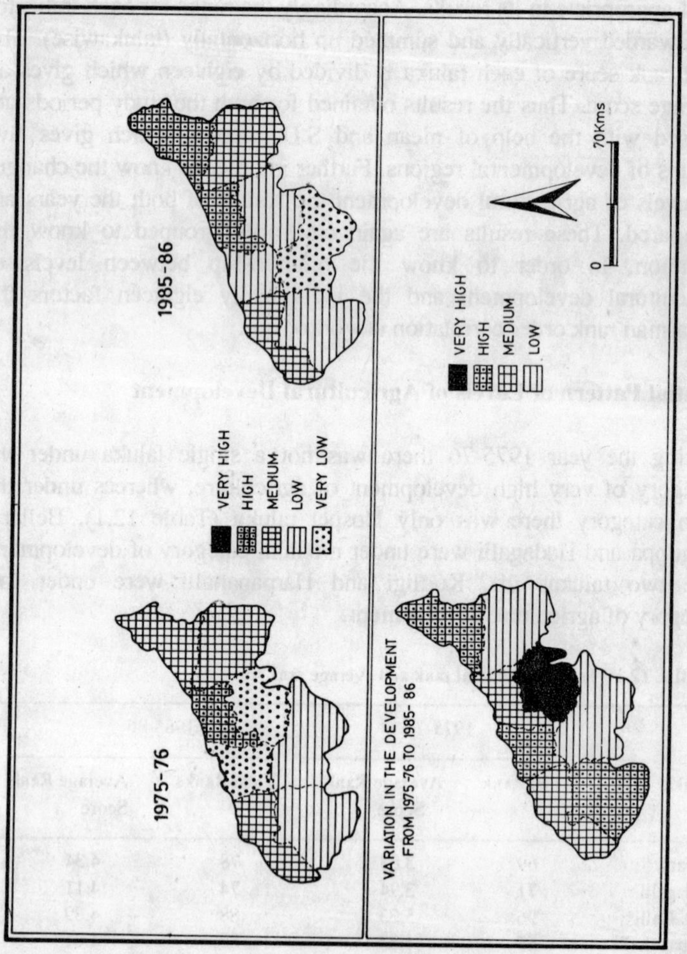

FIG. 12.2: Levels of agricultural development (based on Kendall's rank score method): Bellary district.

TABLE 12.2: Bellary district levels of agricultural development 1975-76 and 1985-86

Category	1975-76		1985-86	
	Index Value	No. and Name of Talukas	Index Value	No. and Name of Talukas
1	2	3	4	5
I. Very High	< 2.65	—	< 3.00	—
II. High	2.65 to 3.51	1 Hospet	3.00 to 3.66	2 Bellary Sirguppa
III. Medium	3.52 to 4.39	3 Bellary Sirguppa Hadagalli	3.67 to 4.34	2 Bellary Hadagalli
IV. Low	4.40 to 5.26	2 Kudligi Harpanahalli	4.35 to 5.01	3 Harpanahalli H.B. Halli Sandur
V. Very Low	> 5.26	2 H.B. Halli Sandur	> 5.01	1 Kudligi
Total		8		8

The remaining two talukas: Sandur, Hagaribommanahalli were under very low development of agricultural group. During this period (1975-76) Hospet taluka had high intensity of irrigation which has a more binding impact on its high development of agriculture. During 1985-86 Sirguppa taluka is added with high group of agricultural development. Hospet, Bellary and Hadagalli talukas which were under medium development of agricultural during 1975-76 have continued in the same group, whereas Sirguppa taluka which was under medium category had changed to high category during 1985-86. Harpanahalli and H.B. Halli talukas had continued as low developed regions whereas Sandur taluka improved its situation from very low to low group. The Kudligi taluka which was under low group during 1975-76 has further come down to very low group during 1985-86. Water in the drier part of the region happens to be a vital factor of agricultural development. This must have had some impact on the agriculture of the Bellary taluka. Therefore, its position as medium development of agriculture did not change.

Changes in Levels of Agricultural Development 1975-76 to 1985-86

The combined impact of the eighteen indicators reveal that Bellary,

Hospet and Kudligi talukas under high category of agricultural developmental have changed. Because of high percentage of area under irrigation and more impact of technological factors in these talukas they are progressing in agriculture (very high change). High change of agricultural development is found in Hadagalli taluka (vide Tables 12.2 and 12.3). The medium change is in Hagaribommanahalli, Harpanahalli and Sirguppa talukas, Sandur taluka alone falls under low change of agricultural development.

TABLE 12.3: Bellary district — Talukawise variation in the indices in the levels of development of agriculture from 1975-76 to 1985-86

Sl. No.	Name of the Taluka	Indices of Agril. Development		Variation in Index Value + or -
		1975-76	1985-86	
1	2	3	4	5
1.	Bellary	3.83	4.34	—0.51
2.	Hadagalli	3.94	4.11	—0.17
3.	H.B. Halli	5.33	4.89	+0.44
4.	Harpanahalli	4.83	4.61	+0.22
5.	Hospet	2.94	3.11	—0.17
6.	Kudligi	4.50	5.36	—0.86
7.	Sandur	5.83	5.06	+0.77
8.	Sirguppa	3.89	3.83	+0.6

TABLE 12.4: Bellary district — Changes in levels of agricultural development

Sl.No.	Category	Index Values	No. and Names of Talukas	
1	2	3	4	
1.	Very High	+ 0.50	1	Sandur
2.	High	+ 0.50 to 0.00	3	Hagaribommanahalli, Harpanahalli and Sirguppa
3.	Medium	0.00 to —0.50	2	Hadagalli and Hospet
4.	Low	— 0.50	2	Bellary and Kudligi
	Total		8	

The study, therefore, suggests that all the talukas which are below the group of high development should be considered for immediate process of agricultural development in order of priority. The development of agriculture is examined with the help of rank order correlation technique. The results reveal that the environmental factors have positive correlation during 1975-76, whereas out of four factors, only the rainfall is negatively correlated during 1985-86. The demographic factors except per capita net sown area to a cultivator are positively correlated during 1975-76. Whereas during 1985-86 except percentage of agricultural workers to the main workers the remaining factors are positively correlated. During 1975-76 the technological factors except percentage area under commercial crops to net sown area and density of livestock other factors are positively correlated with the agricultural development. Whereas during 1985-86 except density of livestock, percentage of electric power consumption and number of pump sets other factors are positively correlated.

13

TRADITIONAL SYSTEMS OF AGRICULTURAL KNOWLEDGE AND CONTEMPORARY AGRICULTURAL DEVELOPMENT IN INDIA

BRIAN J. MURTON

RECENT RESEARCH in India has demonstrated that even where modern inputs have been heavily adopted, farmers have found only some of the package acceptable or practicable : in other words, there has been selective adoption of a system of transferred technologies and practices (Farmer, 1977). As yet very little is known about how farmers integrate new and old technologies and practices into intelligible farming systems, and even less about how modern scientific ideas about agriculture interact with traditional systems of agricultural knowledge. These questions are of considerable importance, because of late it has been recognised that efforts to further increase agricultural productivity in India will depend for the most part on the success with which new agro-technologies and appropriate organisational and institutional arrangements can be incorporated into systems of traditional agriculture (Saint and Coward, 1977).

The purpose of this paper is to direct attention to the desirability of research oriented towards an understanding of traditional systems of agricultural knowledge : that is, the focus will be on cognition as a critical component in understanding Indian agricultural systems. After briefly reviewing the theoretical background of such an approach, a general outline of selected aspects of Indian systems of agricultural knowledge is presented.

Theory and Approach

For nearly 20 years a debate has raged in anthropology over an approach which has been variously labelled "ethnographic description from an emic viewpoint" (Goodenough, 1970), "the ethnographic study of cognitive systems" (Frake, 1962), "cognitive anthropology" (Tyler, 1969) and "ethnoscience" (Sturtevant, 1964). These titles cover a wide range of topics having to do with an approach in ethnography which has focused on discovering how different peoples organise and use their culture as part of an attempt to understand the organising principles underlying behaviour. It is assumed that each people has a unique system for perceiving and organising material phenomena, and that culture is not taken as simply material phenomena, but as a system of knowledge and cognition (Keesing and Keesing, 1971). Sturtevant (1964) puts it this way, a "culture amounts to the sum total of a given society's folk classifications, all that society's ethnoscience, its particular ways of classifying its material and social universe".

A basic premise of the ethnoscience approach in anthropology is that not only do cultures differ from one another in their organisation of material phenomena, but that they differ as well in the kinds of material phenomena they organise, a notion well illustrated by the concept of resources in geography (Zimmerman, 1933; Spoehr, 1956; and Brookfield, 1964). Mention of the resource concept reminds us that while geographers have begun only recently to interest themselves informal ethnoscience, work of a somewhat similar nature has a considerable tradition in the discipline. Historical geographers have long been exhorted to develop "the ability to see the land with the eye of its former inhabitants, from the standpoint of their needs and necessities" (Saner, 1941). In fact, the "earliest methodological discussion along the lines of the present conception of perception in geography" (Saarinen, 1966) was by Kirk, a historical geographer, addressing himself to the question of behaviour of human groups in relation to environment in past times, particularly "as it was observed and thought to be for it is on this behavioural environment that physical features acquire values and potentialities which attract or repel human action" (Kirk, 1952). Of equal significance are the speculations of J.K. Wright, who coined numerous terms, the most popular of which, "geosophy", involves the study of geographical knowledge from any or all points of view. To geography what historiography is to history, it deals with the nature and expression of geographical knowledge both past and present . . . , with what Whittlesey has called "man's sense of terrestrial space" (Wright, 1966).

Thus it is not a new idea in geography that an important aspect of culture is made up of the principles by which people classify their universe: but it is only recently that geographers dealing with contemporary problems in non-Western societies have begun to deal with "the comprehension by indigenous people of what would in the eighteenth century have been called natural history (Brookfield, 1964). Research, closely resembling some of that of cognitive anthropology has been carried out by Porter in East Africa (Knight, 1971, 1974a, 1974b); Knight in East and West Africa (Hansis, 1976); Kirkby in Mexico (Herring, 1975); Hansis in Argentina, Herring in Mexico, and Murton in southern India (Murton, 1973, 1974, 1977a, 1977b).

To date much of this research by geographers has been devoted to the elaboration of terminologies and taxonomies, especially with regard to agriculture. In anthropology, such emphasis upon classification has led some anthropologists to claim that much of the work accomplished under the rubric of ethnoscience has been trivial (Keesing, 1971). However, much of this type of work is a prerequisite for the creation of models of folk science, including systems of agricultural knowledge. Classification is, without doubt, a basic component of both knowledge and thinking, and it becomes a natural starting point for most ethnoscientific studies. However, it does have its limitations and must not be viewed as an all-purpose form for the statement of cognitive systems.

Herring (1975) and Knight (1974b) have suggested how such difficulties can be overcome. Knight says that the next step must be to look at theoretical formulations, empirical relational processes within ethnoscientific systems, and the way in which this theory is used in making behavioural decisions. Herring is more specific in stating that a study of agricultural cognition should take into account the following central questions : (1) what are the principal categories of resource discrimination and what are the bases upon which they are made; (2) how are limited and unpredictable resources combined in such a manner as to make them less limited and more predictable; (3) how are resource-use activities coordinated in cycles of use and access that regulate the timing of the system; and (4) what are the general premises that underpin this body of knowledge and allow the speakers to share similar sets of evaluations and strategies as well as comparable outlooks toward the future. Such a comprehensive approach overcomes the criticism that little research has been undertaken at a theoretical level between terminology and taxonomy, and whole cosmographies (Knight, 1974b). What is required in the Indian case is the delineation of geographical or agricultural theories that translate the ethno-scientific system of Indian society into what Knight has labelled ethno-scientific models (Knight 1974b). These models would contain cognition about environment, rules by which agriculture is undertaken, and information on manual and mental skills, including decision-making devices and logical calculi. They would document cognitive mapping patterns, motivating forces and processes in environment, links to fundamental paradigms of knowledge, and means for coping with risk and uncertainty. Such ethno-scientific models would evoke, as closely as possible, the ability to "see" the environment and agricultural activity through Indian eyes.

Indian Systems of Agricultural Knowledge

I have asserted that attention to understanding Indian agriculture through a cognitive approach may be profitable. Certainly, nobody disagrees that the accumulation and dissemination of knowledge in the form of shared environmental beliefs and rules for production activities is critical for the development and maintenance of complex agricultural systems. To a large extent these culturally defined cognitive patterns are passed through generations by enculturation of children, either informally or in formal education contexts. Such a shared cognitive system serves a number of social functions such as cohesion and control : domains of agricultural production technology are fundamentally critical for society's survival.

Cognitions underlying productive activity are usually characterised by a form of "bounded rationality" (Simon, 1956, 1959). Two connotations of this concept are important in regard to Indian agriculture. First, environmental knowledge and its articulation as agricultural system are internally consistent and, to a large extent, self-fulfilling. Premises upon which actions are taken are reinforced when that action is successful : that the crops yield well justifies the system by which they were cultivated. Second, the concept may usefully be viewed as the local, empirical validity of traditional knowledge. Traced to ultimate origins, folk science may be rooted in religious beliefs or in cosmographies alien to tenets of Western science : these ultimate roots belie, however, the validity and efficacy of environmental knowledge.

With regard to Indian systems of agricultural knowledge, no systematic accounting exists. Yet relevant material has been described in a number of places, and can be used as a general guide.[1] Fundamental to a study of agricultural knowledge is information concerning resources, space and time as environmental events. To begin to elaborate Indian systems of agricultural knowledge systematically I use the four questions posed by Herring as categories into which information can be placed and discussed.

General Premises

Indian folk science is rooted in indigenous religious beliefs and cosmography, and is closely tied to the spiritual fabric of traditional society. The general premises that underlie the folk agricultural science of India include several sets of interlocking principles derived from Indian worldview. The basic concepts which allow people to share similar sets of evaluations and strategies, as well as comparable outlooks towards the future are : *samsara*, the belief in rebirth; *karma*, the belief that actions in one's life determine fortune and status in the next; *dharma*, the definition of what one's behaviour must be in the life status to which one's *karma* has brought one; *moksha*, salvation by release from the wheel of eternal rebirths through the idea of perfect asceticism, of absolute wantlessness; *ahimsa*, the belief in the sacredness of all life (Morris, 1967). The Universe is viewed as consisting of places, each of which may contain gods, men, animals, plants, and things in varying mixtures and proportions (Beals, 1974). In this view there is no separation of Man and Nature and the Supernatural : all have a place, all have a role, all are equally real. However, a sharp distinction is often made between the natural and supernatural worlds. The natural world is divided into ground, mountains, forests, hills, oceans and villages, whereas the supernatural world is divided into separate religious places, such as the loci of Brahma, Siva, Krishna and the Earth. Each of the natural and supernatural places, particularly those regarded as on the earth, contains beings, plants and objects. On the whole this worldview appears to be more ecologically oriented than the traditional Western view.

Perhaps a more important general principle is that within each location all phenomena (people, animals, plants, gods, things) are arranged in hierarchical order. This notion of hierarchy is of critical importance in comprehending systems of knowledge about resources because ranking is a feature of many aspects of resource-use : land categories are ranked; soil types are ranked; crops, even varieties, are ranked. In addition, resource-users themselves are ranked, both in terms of socio-ritual status (*jati* or caste) and in terms of the control and use of land (Dumont, 1970). *Jati*, locally at least, is also an occupational grouping. Thus there is farming *jati*, service *jati*, artisan *jati*, labouring *jati*, pastoral *jati*, trading *jati*, fishing *jati*. Farming *jati* usually ranks high whereas labourers and fishermen are low. But village and local dominance of resources is also a question of the possession of land (Morris, 1967). The system of such interest groups relating to the control of land is neither simple nor uniform throughout India, but it is central to the way in which scarce resources, e.g. irrigation water, are allocated (Frykenberg, 1969, 1977).

Indians also segment the continua of nature : the year is divided into seasons; landscapes are categorised; soils are divided into classes; crop varieties are distinguished. But the Indian mind does not only segment phenomena but arranges them in opposite pairs. Binary opposites are reflected in numerous aspects of Indian life : in the ascribing of heating or cooling attribute to natural phenomena (Beck, 1969), in the polarisation of the year into two major family feast periods (Beck, 1972), in terms of basic geographical polarities, such as north and south, east and west, lowland and upland, centre and periphery; and in terms of opposed landscape types (Murton, 1977). Many of these opposites are mediated by a third term, which reconciles their bipolar tendencies. Furthermore, Indians recognise a few basic substances or elements which underlie the multiplicity of phenomena : earth, water, fire, wind and sky (Ramanujan, 1968). Each substance is identified with a distinctive quality and each is also a process or the embodiment of a principle for action. Each is identified with a range of other phenomena, so that nature is made intelligible and coherent.

Of equal importance is traditional South Asian understanding of time, which is not thought of in linear progression, but in terms of cycles, which repeat themselves endlessly (Embree, 1968). Time moves in cycles within cycles — days within astral periods within months within seasons within years within eras of approximately 300 million years. Another characteristic of Indian time is that it has no beginning or ending : there is only endless repetition of cycles. Finally, nothing is more indicative of the all-encompassing nature of time than the gods, no less than men, who are part of the eternal cycle. There is no dividing line between the human and the divine, there is a continuum between man and nature and between man and the gods, because all are bound in the movement of time.

Co-ordination of Agricultural Activities

The key operation which ties together the events in the chain which links environment with resource potential is cognition especially appraisals of the environment, which, along with overt resource behaviours, constitute part of the Indian definition of "proper agricultural practices". The coordination of agricultural activities has both temporal and spatial aspects. Traditional systems of dividing time are common to all parts of India. Three methods of computing yearly cycles are in use (solar, junar, *fasli* or revenue year), but in terms of agriculture the solar year which begins in mid-April is the most important as its beginning more or less coincides with the thundershowers of April which herald the coming monsoon, the beginning of the agricultural season.

The exact number and dates of seasons in the year vary from place to place, but there are usually six, of two months each. As important as recognised seasons are sowing periods which are highly localised, reflecting appraisals of the moisture in environment in relationship to particular crops. But perhaps the most important devices for regulating the timing of agricultural operations are astral periods (*nakshatra*). These are formulated in terms of the relationships of the sun, the 12 signs of the zodiac and 27 stars. There are 27 periods, to each of which are assigned agricultural operations and tasks.

If times is organised into years, months, seasons and astral periods, space is organised into a number of land-soil-crop associations. Throughout India certain crops or crop associations are related to certain land and soil types. These inter-relationships are models which farmers use to guide them in their agricultural activities : indeed, they are culturally prescribed ways of doing things.

However, Indian agricultural knowledge is not confined to a sum of unconnected parts. Environmental appraisal, as well as agriculture, requires that terms be linked in webs of associations : sorghum intersects with certain land types, soil/water conditions, insect pests; rice with others; wheat with yet others. Different seasons intersect with the availability of different forage plants, different weather conditions intersect with different diseases, and different terrains with different plant associations. Thousands of these simple linkages constitute the building blocks of traditional Indian agricultural strategy : the ways in which resources are combined and timed.

Resource Predictability

Time and space are organised into sequences and segments, linked into webs of association. A wealth of knowledge associated with these makes the somewhat limited and unpredictable resources of India less limited and more predictable. This accumulation of experience is formalised in the form of folk songs, sayings and proverbs (Gover, 1871; Gupta, 1963; Balasubramanian and Gopalkrishna, 1953; Balasubramanian and Rao, 1950). The months of the year, the seasons, and especially the astral periods, all have associated with them a vast array of pithy statements about every aspect of tillage and agricultural operations. In the past, and even today, agricultural operations were guided by such statements, especially those relating to rainfall and the appropriate timing of cultivation practices. Much of this information about rainfall is contained in agricultural almanacs which predict in detail the expectation of rainfall based upon the position of the stars in relationship to the moon. Indian farmers have also developed devices to make their uncertain environment determinate and knowable, most of which involve a belief in the use of natural phenomena to predict the occurrence or non-occurrence of rainfall. Land and soil types also have sayings and proverbs associated with them, as do most agricultural practices, crops and livestock. Proverbs also allude to the knowledge of agriculture held by some *jati* and the lack of it by others.

Categories of Resource Discrimination

The principal categories of resource discrimination in India include soil, slope, water, climate, time, space, plants, animals, and fish, all aspects of environmental elements. Such terminological domains are further elaborated in taxonomies. Indians have classifications of landscape morphology, land types, of soils, of vegetation, of crops, of macro-climatic and micro-climatic resources. These are related to the general principles discussed earlier, and although each can be isolated and reviewed separately, in reality they make up a complex of intersecting, interconnected entities and events.

Agricultural classifications reflect local characteristics of the biophysical environment. Throughout the region land is initially classified as either "wet" or "dry". This distinction between irrigated and rainfed land is of major importance because distinctive types of management and conservation practices are related to wet and dry land. In the Ganga Valley and delta areas land is further classified in terms of elevation above water level. Soils are most often initially classified according to their colour, e.g. "red" or "black", and then in terms of very fine textural discriminations, as well as distance from settlement sites. Crop varieties are classified in terms of obvious seed colour, height, seedhead shape, duration, taste, and other visible characteristics.

In summary, considerable empirical knowledge of the physical world, and taxonomies of a fair degree of sophistication exist. This resource knowledge is tied to the fundamental paradigms of knowledge in Indian society through the set of general principles. These paradigms ultimately cross the boundary between science and religion, even if day-to-day forces in the environment do not. In fact, to dichotomise between science and religion is undoubtedly erroneous for many Indian farmers, even today : a simple pragmatic solution to a problem may well be an appropriate ceremony or prayer.

Conclusion

Today, major changes have begun to manifest themselves with regard to Indian farmer's knowledge and understanding of their environment. More than one hundred years of contact with Western scientific ideas, and the acceptance of many of the basic premises of Western science by the educated elite at least, has led to apparent widespread changes in beliefs about the natural world. India has an elaborate and sophisticated scientific and technological community, which is involved in the education process and basic research. These are the people and institutions which are attempting to change existing agricultural technologies and practices into something new, and that new something is a rational, or at least rationalistic, scientific system of agricultural management.

Thus India is the theatre of an indescribably complex interaction between the forces of modern science and technology and of an age-old metaphysical tradition. Despite numerous advances, in the villages custom is a king, not yet and not easily dethroned (though increasingly tottering), and his laws are enshrined in the multitude of pithy proverbs which, with religious songs, form a remarkable folk system of knowledge. That this has survived to the extent that it is fortunate because it is beginning to be recognised that indigenous knowledge has something to contribute to efforts to develop agricultural systems. To date, in its arrogance and ethnocentricity Western science has only too often pushed aside the wealth of folk knowledge and experience that is embodied in traditional agricultural systems. But in an age of potential shortages of energy to fuel the new agro-technologies it has become doubly important to have some understanding of systems of traditional agricultural knowledge. In particular, it is being recognised that agricultural systems, which have existed for as long as those of India, represent at least partially successful solutions to specific sets of man-environment situations, and that the systems of knowledge associated with these complexes may contain valuable information even if at the moment it appears somewhat anachronistic to those trained in Western science.

NOTE

1. Such sources include the periodicals *Man in India* and *Folklore*, various agricultural journals, such as the *Madras Agricultural Journal*, e.g. C. Balasubramanian and R. Gopalkrishnan, "Agricultural Folklore in Malabar", *Madras Agricultural Journal*, Vol. 40 (1953), pp. 55-59; the District Gazetteers of British India (see Henry Scholberg, *The District Gazetteers of British India : A Bibliography* (Zug, Switzerland : Inter Documentation Company, 1970), books such as George A Grierson, *Bihar Peasant Life* (Calcutta : Bengal Secretariat Press, 1885), and numerous books, monographs and articles dealing with religion and philosophy.

REFERENCES

Balasubramanian, C. and Gopalkrishnan, R., 1953 : "Agricultural Folklore in Malabar", *The Madras Agricultural Journal*, Vol. 40, pp. 55-59.

Balasubramanian, C. and Narasinga Rao, M. B. V., 1950 : "Meteorology and Agriculture", *The Madras Agricultural Journal*, Vol. 37, pp. 239-41.

Beals, A., 1974 : *Village Life in South India : Cultural Design and Environmental Variation*, Chicago : Aldine Publishing Company, p. 30.

Beck, Brenda E. F., 1969 : "Colour and Heat in South Indian Ritual", *Man*, Vol. 4, pp. 553-72.

———, 1972 : *Peasant Society in Konku : A Study of Right and Left Subcastes in South India* (Vancouver : University of British Columbia Press), pp. 52-56.

Brookfield, H. C., 1964 : "Questions on the Human Frontiers of Geography", *Economic*

Geography, Vol. 40, p. 287.

Dumont, Louis, 1970 : *Homo Hierarchus. The Caste System and its Implications,* Chicago : The University of Chicago Press, contains a comprehensive discussion of caste.

Embree, Ainslie, T., 1968 : "Tradition and Modernisation in India Synthesis or Encapsulisation", in Ward Morehouse, ed., *Science and the Human Condition in India and Pakistan* (New York : The Rockefeller University Press), pp. 29-38.

Frykenberg, Robert E., ed., 1969 : *Land Control and Social Structure in Indian History* (Madison, Wisconsin : University of Wisconsin Press).

———, ed., 1977 : *Land Tenure and Peasant in South Asia,* New Delhi : Orient Longman.

Farmer, B. H., 1977 : "Technology and Change in Rice Growing Areas".

Farmer, B. H., ed.. 1977 : *Green Revolution? Technology and Change in Rice-Growing Areas of Tamil Nadu and Sri Lanka* (Boulder, Colorado : Westview Press), p. 4.

Frake, C. O., 1962 : 'The Ethnographic Study of Cognitive Systems", in Gladwin, T. and Sturtevant W.C., eds., *Anthropology and Human Behaviour* (Washington, D.C. ; Anthropological Society of Washington), pp. 72-85.

Goodenough, W. H., 1970 : *Description and Comparison in Cultural Anthropology* (Chicago : Aldine Publishing Company, 1970), p. 111.

Gover, C. E., 1871 : *The Folk Songs of Southern India,* Madras : Higginbotham and Co.

Gupta, S. S., ed., 1963 : *Rain in Indian Life and Lore* (Calcutta : Indian Publications).

Hansis, R. A., 1976 : *Ethnogeography and Science : Viticulture in Argentina,* unpublished Ph.D. Dissertation Pennsylvania State University.

Herring, K. J., 1975 : "New Techniques for the Construction of Ethno-sciences", Paper presented at Annual Meetings of the Association of American Geographers, Milwaukee.

Keesing, R. M. and Keesing, F. M., 1971 : *New Perspectives in Cultural Anthropology* (New York : Holt, Rhinehart and Winston), pp. 19-28.

Keesing, R. M., 1977: "Paradigms Lost: The New Ethnography and the New Linguistics", unpublished manuscript.

Kirk, W., 1952: "Historical Geography and the Concept of the Behavioural Environment," *Indian Geographical Journal*: Silver Jubilee Edition, 1951, Madras: Indian Geographical Society, p. 159.

Knight, C. G., 1971: "Ethnogeography and Change", *The Journal of Geography*, Vol. 70, pp. 47-51.

———, 1974a: *Ecology and Change: Rural Modernisation in an African Community*, New York: Academic Press.

———, 1974b: "Ethnoscience: A Cognitive Approach to African Agriculture", Paper presented at the S.S.R.C. Conference on Environmental and Spatial Cognition in Africa, Los Angeles.

Kirkby, A. V., 1973: "The Use of Land and Water Resources in the Past and Present Valley of Oaxaca, Mexico", *Memoirs of the Museum of Anthropology* No. 5 (Ann Arbor, Michigan: University of Michigan).

———, 1974: "Individual and Community Responses to Rainfall Variability in Oaxaca, Mexico", in White, G. F. ed., *Natural Hazards. Local National, Global* (New York: Oxford University Press), pp. 119-128.

Morris D. Morris, 1967: "Values as an Obstacle to Economic Growth in South Asia: An Historical Survey", *Journal of Economic History*, Vol. 27, pp. 588-607.

Murton, B. J., 1973: "Folk Classifications of Cultivated Land and Ecology in Southern India", *Proceedings of the Association of American Geographers*, Vol. 5, pp. 199-202.

———, 1975: "Territorial and Resource Structure of Tamil Villages in 1800: Implications for Settlement Evolution", in Brockie, W., Leheron, R. and Stokes, E., eds., *Proceedings of the International Geographical Union Regional Conference and Eighth New Zealand Geography Conference, Palmerston North*, 1974 (Hamilton: New Zealand Geographical Society), pp. 119-26.

———, 1977a: "Land and Class: Cultural, Social, and Biophysical Integration in Interior Tamil Nadu in the Late Eighteenth Century", in Frykenberg, ed., *Land Tenure and Peasant in South Asia* (Delhi: Orient Longman), pp. 81-99.

———, 1977b: "Changing Land Categories in Interior Tamil Nadu, 1750-1850: Propositions and Implications for the Agricultural System", in Eidt, R. C.; K. N. Singh, and P. B. Rana Singh, eds., *Man, Settlement and Culture, Festschrift to Prof. R. L. Singh* (New Delhi: Kalyani Publishers), pp. 31-40.

———, 1977: "Changing Landscape Categories in Tamil Nadu, 1750-1850: Propositions and Implications for the Agricultural System", in Robert C. Eidt, Kashi N. Singh and Rana P. B. Singh, *Man, Culture and Settlement. Festschrift to Prof. R. L. Singh* (Delhi: Kalyani Publishing House), pp. 31-40.

Porter, P. W., 1963: "Suk Views on Suk Environments", Paper presented at Annual Meetings of the Association of American Geographers, Denver.

———, 1965: "Environmental Potentials and Economic Opportunities — Background for Cultural Adaptation", *American Anthropologist*, Vol 67, pp. 409-20.

———, : "Pokot Ideas Concerning Soil", *Mimeographed*, no date.

Ramanujan, A. K., 1968 : "Form in Classical Tamil Poetry", Paper presented at the Meeting of the Society for South Indian Studies, Austin, University of Texas.

Saarinen, T. F., 1966 : *Perception of Drought Hazard on the Great Plains*, Research Paper No. 106 (Chicago : University of Chicago Department of Geography), p. 32.

Saint, W. S. and Coward, E. W., 1977 : Agriculture and Behavioural Science : Emerging Orientations", *Science*, Vol. 197, p. 733.

Sauer, C. O., 1941 : "Foreword to Historical Geography", *Annals of the Association of American Geographers*, Vol. 31, pp. 1-24.

Simon, H. A., 1956 : "Rational Choice and the Structure of the Environment", *Psychological Review*, Vol. 63, pp. 129-38.

———, 1959 : "Theories of Decision Making in Economics and Behavioural Science", *American Economic Review*, Vol. 49, pp. 253-83.

Spoehr, A., 1956 : "Cultural Differences in the Interpretation of Natural Resources", in Thomas, W. L. ed., *Man's Role in Changing the Face of the Earth* (Chicago : The University in Chicago Press), pp. 93-102.

Sturtevant, W. C., 1964 : "Studies in Ethnoscience", *American Anthropologist*, Vol. 66, Part 2, pp. 99-131.

Tyler, S. A., ed., 1969 : *Cognitive Anthropology* (New York : Holt, Rhinehart and Winston).

Zimmerman, E. R., 1933 : *World Resources and Industries* (New York : Harper and Brothers), pp. 3-20.

Wright, J. K., 1966 : *Human Nature in Geography : Fourteen Papers, 1925-1965* (Cambridge, Massachusetts : Harvard University Press), p. 83.

14

AGRICULTURAL DEVELOPMENT AND ENVIRONMENTAL ISSUES

SURENDER SINGH

AGRICULTURE IS undoubtedly one of the most important economic activities of man. It provides not only the food to the teaming billions of the world but also a number of industrial raw materials. In India, the importance of this activity is all the more greater as nearly 64 per cent of the total population is even now dependent upon farming. About 47 per cent of the total land area of the country is under plough and more than 45 per cent of the national income is derived from this sector of the national economy.

The large population of India puts an ever-increasing pressure on the agricultural resources. It has caused frequent shortages of foodstuffs and other agricultural products in the country. The population of India which was approximately 361 million in 1951, increased to more than 685 million in 1981. It is estimated that the population of India at the time of 1991 census will be more than 850 million and it is likely to reach the one billion mark by the turn of the present century. As a consequence of this rapid growth of population, the requirement of foodgrains alone is likely to be about 230 million tonnes by the year 2001 A.D.

The production of foodgrains in the country was less than 51 million tonnes in 1950-51, and by 1984-85 it could be raised to about 155 million tonnes. Incidentally, this amount of foodgrains production is considered by many scholars as exceeding the inherent capacity of our soil resource under normal intensity of landuse. Thus the question facing the modern agricultural scientists is "how to increase the production of foodgrains by about 50 per cent over a short span of time of about 10 to 15 years"? At this stage it is relevant to note that as the requirement of foodgrains is increasing, the demand for other agricultural products is also rising. Thus the seemingly apparent solution involving increasing the area under foodgrain crops at the cost of other crops, cannot be relied upon too heavily. We have already experienced some effects of this policy in the form of reduced availability of pulses and oil seeds. During the past about four decades we increased production of foodgrains by nearly 300 per cent thereby increasing the availability of foodgrains vis-a-vis the population of the country. However, during the same period, the per capita availability of oil seeds, edible oils and pulses has declined alarmingly, so that some of these items are becoming increasingly scarce in the diet of the poor of the country. This has, of course, been primarily due to the excessive emphasis of our policy on increasing the production of foodgrains. Thus what we achieved in real terms was better capability to feed people, perhaps, by bringing nutrition levels down.

There are only two options to follow for increasing the production of foodgrains to match the ever-growing demand. The first option is to increase the area under cultivation. As observed earlier, the area under foodcrops cannot be increased by diverting cultivated land from other crops. The other way to increase this area is by bringing additional land under plough. This possibility also has very little scope. In India, the possibility of converting forest and other lands into cultivated land are meagre. Our natural vegetation cover is already alarmingly low. At the same time, large areas of cultivated land are being taken away by other uses such as urbanisation, industrialisation and transport. Thousands of hectares of land are annually encroached upon by expanding cities, new and expanding industrial estates and highways. More often than not, such lands are of prime importance from the agricultural point of view. Thus the loss of agricultural land to other uses usually more than neutralises any advantages accruing through conversion of marginal lands into cultivable lands. Therefore, in real terms, the possibility of increasing the cultivated area of the country, holds extremely limited promise.

The second way in which we can increase the production of foodgrains is by increasing the per hectare yields. In fact the per hectare yields of many of our foodcrops are still very low, inspite of the fact that we have great achievements of our credit in this field. Much of the increase in the production of foodgrains during the past four decades has been a result of increased productivity. However, the per hectare yields of the major foodgrains such as rice and wheat are quite low when compared to the yields of the same crops in many of the other countries. Productivity of Chinese agriculture in respect of wheat and rice is almost double our productivity in these crops. It is mainly for this reason that China can feed a population which is one and a half times the population of India, from the production of a cultivated area which is lesser than the cultivated area in our country.

Keeping in view the productivity levels in countries like China, there is ample scope of increasing the per hectare yields of most of the foodcrops in India. This can be achieved through adopting suitable technology, better management of land and other resources and by developing the infrastructure of agriculture in India. In fact the very adoption of modern technology will call for numerous changes in the infrastructural component. However, this is not major theme of discussion in the present analysis.

Adoption of the developed technology has shown positive results in respect of agricultural efficiency and productivity the world over and India is not an exception here. It is mainly through the adoption of the developed technology that we have increased the production of foodgrains in our country by nearly 300 per cent within a period of less than four decades. The "Green Revolution" has been possible only through the adoption of the technology of the high yielding varieties of seeds, supported by the use of chemical fertilisers, irrigation and other modern inputs. Thus the key solution to the future supply of foodgrains is the application of the technology of the "Green Revolution" in areas which have so far not benefited from this and adoption of some more advanced forms of technology all over the country. However, all this is going to have its toll in the form of high economics and environmental costs. The more important components of the existing technology in the field of agriculture and some of the environmental problems associated with these are enumerated in the following paragraphs.

Bringing Marginal Lands under Plough

The cultivated area in the world as a whole amounts to approximately 11 per cent of the total land area of the earth, while the permanent pastures and forests and woodlands account for about 24 per cent and 31 per cent, respectively, of the total. The 11 per cent area of the total which is presently under cultivation in the world, represents about 50 per cent of the potentially cultivable area of the earth. The proportion of the potentially cultivable area which is under plough presently, amounts to about 70 per cent in the developing countries, while this proportion in the southeastern Asian countries is as high as 92 per cent. In India also this proportion is quite high as is evident from the fact that in states like Punjab and Haryana, more than 84 per cent of their geographical area is under cultivation. Many of the other states including Uttar Pradesh and Maharashtra have almost 60 per cent of their total area under cultivation. West Bengal has more than 62 per cent of its total area under plough and even in a semi-arid region like Gujarat state, this proportion is more than 51 per cent. Most of the remaining potential cultivable land in the country lies in areas which are either too dry or are mountainous or where the land and soils require a good deal of management effort before these areas can be brought under plough.

The area under forests in our country is already less than 15 per cent of the total, as against a desirable proportion of about 33 per cent. Under the prevailing circumstances any meaningful increase in the area under cultivation can be either through bringing marginal lands under plough by way of huge economic and technological investments or through encroachments upon forest and woodland areas. This later course has been resorted to more frequently in the past as is quite logical in a country inhabited by farmers having limited financial capacities to improve and bring under plough the marginal lands.

The decreasing area under forests has a telling effect upon the quality of environment and the ecological conditions in the country. A number of areas in India are already experiencing extremely high rates of soil erosion, resultant land degradation and floods and droughts. Most of the rivers flowing over Himalayan ranges carry huge amounts of debris contributed by soil erosion on the devegetated slopes. According to a number of studies, the Thar Desert is expanding east and northeastwards due to declining plant cover and overutilisation of soils. The formation of gullies and ravines in much of the central part of the country is partly due to deforestation. The total water-eroded area of the country amounts to more than 73 million hectares and the total non-forest land seriously affected comes to more than 93 million hectares. The total cultivated area suffering from erosion in India works out to 41.5 hectares. Presently the country is losing about 8000 hectares annually to ravines.

As the marginal lands such as semi-arid areas and the foothills of the mountainous and hilly regions are converted into cultivate lands, the whole ecological system of such areas is altered. The soil layer which must have taken millions of years in developing and accumulating under various types of vegetation, is washed away within no time and the whole area is rendered unsuitable not only to cultivation, but also to seemingly less intensive uses such as cattle grazing. Today hardly any soil is left on the hill slopes in many parts of Himachal Pradesh and Uttar Pradesh Himalayas. In quite a few pockets, the deposition of coarse debris brought by land slides and mud-flows etc. covers the productive valley lands thus rendering these areas uncultivable. All this is mainly a result of human activities such as deforestation, over-grazing and cultivation over steep slopes. Thus the benefits of increasing the area under plough are usually more than neutralised by the environmental costs resulting from such moves.

Besides this, there is another hurdle in the way to increase the area under cultivation. Currently a substantial amount of cultivable land is being lost to other types of uses, prominent among which are urban residential, industrial and transport uses. A reference to loss of good quality agricultural land to such uses has already been made. If these losses are added to the losses due to soil erosion and land degradation, they more than balance the rate at which new land is being brought under plough. The diversion of good quality, fertile valley lands is going on in most parts of the northern plains of India, thus restricting the scope of any increase in the cultivated area.

High Intensity of Landuse

Under the prevailing conditions of scarcity of cultivable land and the growing needs of agricultural produce, the most logical wayout is to use the available resources more intensively. The natural consequence of this need is the increased emphasis on multiple cropping, irrigated farming and cultivation on marginal lands. The intensity of agricultural landuse has increased considerably almost the world over. It has helped in raising the production of foodstuffs, but at the same time has caused a number of environmental problems. The most serious problem emanating out of higher intensity of landuse is the declining soil fertility and land degradation. Under normal conditions, intensities of landuse higher than a certain limit cannot be sustained. Under continuous use the soil tends to become increasingly poor in plant nutrients. Soil fertility is depleted not only by a larger number of crops per unit of time but also by over-emphasis on food crops. As increasingly higher proportions of cultivable land are used for growing foodgrains, the possibility of land being left fallow is marginalised. At the same time, green manuring crops and other crops helping in replenishing soil fertility also get lesser area. This precludes the chances of soil fertility being retained at a sufficiently high level. Mono-culture of wheat during rabi season and that of rice during kharif season in western parts of Ganga Plains has reduced soil fertility considerably. Higher intensity of landuse is frequently achieved through mechanisation. This reduces the number of farm animals needed to work a given area of land. Undoubtedly it reduces the need for area required for growing fodder crops and thus releases more land for growing foodgrains. However, at the same time it reduces the availability of manure for farmland, thereby adding to the problem of declining fertility of soils. In the traditional system of Indian farming the animals have been performing an important function in recycling soil nutrients from the farm waste materials like the crop residues, e.g. the cheff of wheat stalks. Lack of animal has tended to prolong the period of recycling of these nutrients in the natural course.

Mechanisation

Mechanisation of agriculture in India has helped tremendously in increasing productivity. It has also helped in bringing marginal lands under plough. Also, many of the farming operations have been made more efficient and rapid thereby saving a lot of foodgrains from vagaries of weather elements. However, mechanisation of farming operations extracts its own costs.

Firstly, in a society where human labour is cheap and abundantly available, use of machines tends to increase the cost of production. It is more so when the energy sources such as oil entail heavy costs as at present. The high cost of production of foodgrains leads to higher market prices and inspite of higher production of grains in the country, the hunger cannot be removed in a poor labour intensive society where the masses do not have a sufficiently high purchasing capacity. However, the economic consequences of mechanisation are not to be analysed at length here. The major emphasis of the present analysis is on the environmental consequences of modernisation of farming operations.

A greater emphasis on use of machinery in agriculture leads to some environmental costs too. The most noteworthy impact of use of fossil fuels operated technology, as the farming technology at large, is the addition of carbon and sulphur compounds into the air. It is another matter that the concentration of these pollutents in the rural environments is not too high at present and in the near future it is not likely to pose a serious pollution threat.

A more serious consequence of mechanisation is on the soil environment. Use of heavy machines like tractors and combine harvestors causes compaction of soil and it interferes with the soil texture and structure. The net impact of these changes is the reduced soil porosity and water holding capacity. Reduced porosity checks the rate of infiltration of water and leads to a higher proportion of the rainfall beings lost as run-off. One of the consequences of this is a higher rate of soil erosion and another is a lower rate of recharge of groundwater reservoir. Compaction retards soil drainage, and reduces the depth of plant rooting zone. According to some estimates, during early 1970s crop yield reductions due to compaction of soils varied from one per cent in northern USA to about 10 per cent in southern USA. By the year 1980, the losses in monetary terms had trebled.

Mechanisation in India has led to human tragedies also. During the harvesting season, there are a number of accidents involving threshing machines and there are losses of limbs as well as of life. This is more a result of faulty design of the threshing machines. The loss of organic manure supplied by the farm animals, as they are replaced by machines has already been mentioned. Thus in real terms, the cost of mechanisation is much higher than what is apparent in money terms.

Irrigation

Another important means to increase agricultural productivity is irrigated farming. Irrigation has been one of the most important infrastructural components supporting green revolution in India. Through artificial watering of crops not only the yield can be increased but also the cropped area can be multiplied by bringing additional land under multiple cropping. The gross irrigated area in India amounts to slightly more than 30 per cent of the gross sown area and this amounts to about half of the total potential irrigated area. Irrigation in our country is undertaken through wells/tubewells and canals primarily, the tanks accounting for less than 10 per cent of the net irrigated area. The proportion of the area irrigated by means other than canals and wells is declining.

Canals use the water from the streams and quite a few of these have been taken out of rivers without constructing any dams etc. In addition to these canals, there are the canals which are taken out after constructing dams across the rivers. This leads to a number of environmental and other complications. The most significant problem emanating out of this is the change in the hydrological balance of the downstream areas in the river valleys. Besides this, the accumulation of water behind the dams leads to not only the problems of disturbing the isostatic balance but also to the submergence of precious cultivable areas in the valley bottoms. Quite frequently large farming populations have to be shifted at the time of the construction of these dams and this uprooting of large farming populations leads to untold social and economic costs. Many a time the affected people resort to agitation opposing the construction of dams. Tehri Dam and Narmada Sagar projects are only two among so many projects which have become controversial on account of environmental and socio-economic considerations.

In addition to the general problems related to the irrigation projects as outlined here, there are some problems which have to be faced by the farmers themselves. They are almost purely ecological in nature. Intensive canal irrigation has led to increasing soil salinity and alkalinity in many parts of the world over. Intensively canal irrigated regions in Uttar Pradesh and Punjab have shown an increasing trend in the area covered by alkaline and saline soils. The increasing salinity is caused primarily by accumulation of salts brought over a long time in solution in the canal waters. Alkalinity on the other hand is caused mainly in areas of poor soil drainage and saturation of soils leading to alkaline deposits on the soil surface is a common problem in such regions. Soils having a poor drainage are not well suited for intensive irrigation and if the irrigation is continued for long, the soils are rendered useless from the cropping point of view. Quite frequently heavy doses of irrigation might lead to the problem of water-logging also. This problem is quite acute in parts of the Rajasthan Canal command area. In this area a thick layer or pan of gypsum is found at a rather shallow depth and according to some studies about 8 per cent of the command area of the project is already water-logged. The present rate of rise in the water table in this region is about 2 ft. per year and at this rate, about a quarter of the command area of the project will be water-logged in a few years. The problem of water-logging is faced in many parts of Punjab and Haryana also where the water table is rising very rapidly along the banks of the unpaved canals.

Although much of the increase in the irrigated area in India has been achieved through expansion of the canal systems, the dependence upon the wells and tubewells has not decreased. This means of irrigation accounts for about 46 per cent of the irrigated area in the country. In fact the scope of expanding the canal systems in many parts of the country is rather limited due to lack of surface water resources which precludes large scale development of canals. Most of such regions receive too little amount of rainfall to provide water for canals. The minor irrigation schemes seem to be the most suitable alternative in such areas. Most of these regions are included in the zone of dry land farming. In all such regions the irrigation system can be augmented only through more intensive use of groundwater resources. Hence, the importance of wells and tubewells is not likely to decline in near future also.

Large scale mining of ground water in such areas receiving low rainfall poses serious problems in the long run. In all such regions the rate of accrual to the ground water body is very low and mining of water at a rate higher than this rate may cause depletion of the ground water reservoir. The earliest sign of this problem is the falling water table, as is the case in many parts of Haryana, Gujarat and coastal parts of Tamil Nadu and Orissa. Many a time the water table falls so low that the water from the wells or tubewells cannot be lifted by ordinary means such as animal power or the ordinary pumping sets installed at the ground level. A large number of tubewells and wells in the above noted areas have to be deepened periodically and it is feared that in the near future at least some of the wells might go dry or the water table will fall so low that the cost of lifting it will be very high.

Rapid fall of water table and depletion of the ground water reservoir may pose a number of other difficulties besides the high cost of irrigation. In coastal districts of Gujarat, hundreds of wells have been rendered unsuitable for irrigation or other intended uses during the recent years. Here the rapid fall of water table has caused landward penetration of saline water from the sea. As this water was used for irrigation, crusts of salt were deposited on the soil surface thus rendering it unsuitable for cultivation. Beginnings of a similar problem have been observed in coastal Tamil Nadu and Orissa also. In semi-arid region of Haryana where tubewell and well density is high, the water table has fallen very rapidly. Here there are lenses of brackish water at depth below aquifers containing fresh water. As the top aquifer is depleted, the wells or tubewells have to be bored down up to the brackish water aquifers, thus rendering many of the wells unsuitable for supplying drinking water and often even irrigation water.

Human tragedies have been reported from Karnal district of Haryana where lowering of water table necessitated installation of pumping sets in very deep hollows (dry wells). These hollows become accumulation points of heavier gases such as carbon-di-oxide and a number of persons who descended into these dry wells to repair the pumping sets, died of lack of oxygen in the pits. If the horizontal migration of saline water is not impeded by geological structure, the saline water may penetrate even to such interior locations. Drinking water may also become scarce in such inland locations also as has happened in many coastal pockets. The supply of drinking water may be affected by the falling water table as the handpumps used for lifting it might start becoming useless as the water will have to be lifted from increasing depths.

Rapid fall of water table may lead to land sinking also. In fact the rate of fall of water table in some areas is so high that possibilities of land sinking are magnified. As per some recent studies, Chandigarh region has an alarmingly high rate of fall of water table — more than one metre per year. Any type of catastrophes might be expected in such regions.

Intensive use of irrigation necessitates modifications in the cropping pattern also. In most parts of Punjab, Haryana and western Uttar Pradesh, the area as well as the production of pulses and oilseeds has declined rather dramatically ever since the increased emphasis on irrigation in the wake of introduction of high yielding varieties. Most of the high yielding varieties of seeds are irrigation intensive, and pulses and oilseeds which were earlier grown as mix crops with wheat and gram etc. under very little or no irrigation, cannot be grown with these varieties. This has been at least partly responsible for the reduced availability of these agricultural products in the country.

Higher degree of irrigation leads to a higher incidence of pests and plant diseases also. Pests pose a serious threat to crops over vast areas of irrigated farming as and when they spread. Incidence of diseases like malaria, filariasis and liver fluke infections has been observed to have increased after the introduction of irrigation system in many parts of the world. With introduction of irrigation, the whole ecological conditions of the areas concerned are altered and quite frequently these alterations are harmful. All this shows that irrigation, though apparently a boon to the farming, may lead to serious environmental consequences, if introduced without taking due precautions or if over used.

High Yielding Varieties of Seeds

High yielding varieties of seeds have been a revolutionary development in the field of agriculture during the twentieth century. Use of such varieties of seeds, especially of wheat and rice, has increased the production of food grains in India and it has been one of the supporting pillars of the green revolution. These varieties have become increasingly popular during the recent decades and the area planted with these varieties of seeds has registered a steady growth. The total area under the high yielding varieties in India was less than 20 lakh hectares in 1960-61 and by 1986-87 it had increased to more than 540 lakh hectares. In fact much of the increased production of foodgrains as well the higher productivity is a result of this revolution in the field of agricultural technology.

Ecologically speaking the high yielding varieties have a very high efficiency of nutrient absorption and food synthesis. They take up large quantities of nutrients from the soil which forms the basis of higher yields per hectare. Therefore, the depleting soil fertility is the foremost consequence of the use of these seeds. Unless higher doses of nutrients are provided to the farmland in form of organic manures or the chemical fertilisers, the high yields of crops cannot be sustained even with the help of these varieties of seeds. The water requirements of these varieties are also many times greater than those of the traditional varieties. This results in a much greater stress upon both the surface and the ground water resources of the areas concerned. In fact the entire technology becomes a highly costly affair not only from an economic point of view but also from an ecological standpoint. It is the higher requirements of inputs like water, chemical fertilisers and the consequent higher monetary inputs which have tended to keep the green revolution restricted to only a few areas of India. Incidentally all the areas which participated in this revolution are the ones where the infrastructure had already developed and large monetary inputs in irrigation etc. did not have to be made.

Most of the high yielding varieties of seeds have been developed from some imported varieties which are not adapted well to the local environmental conditions; at least not as well adapted as the local varieties are. This makes the new varieties more prone to common pests and plant diseases. In addition, they might become susceptible to some new plant pests or diseases also against which the local varieties have a very high resistance. Also, if a pest or some disease affects one plot of crop, the same malady is likely to spread over a much larger area. This is so as most of the high yielding varieties have been developed from a few common genetic strains.

Introduction of a high yielding variety of seeds of a particular crop tends to make such a crop more profitable in terms of the total income per unit of area per unit of time. The farmer tends to grow such a crop to almost the complete exclusion of all other crops, leading to the emergence of a monocultural cropping pattern. This also adds to the problem of declining fertility of soils. At the same time it increases the probabilities of crop failures due to spread of diseases or pests. Changes in the cropping pattern may lead to alterations in the dietary habits of the people. This may lead to health problems among the masses. One of the examples of this type of a change has been observed in the Chambal Valley Project command area. Here with the introduction of irrigated rice farming, the traditional millets dominated diet of the people was replaced by a rice dominated diet. As a consequence the incidence of disorders of the digestive system increased in the whole area. Similar changes in the dietary habits have been observed in parts of Punjab and Haryana region where millets, pulses and oilseeds have been almost totally excluded from the prevailing cropping pattern, and the proportion of these food items has decreased in the diet of the people.

Chemical Fertilisers

Use of high yielding varieties of seeds as well as a higher intensity of landuse lead to a rapid depletion of soil fertility as the rate of replenishment of nutrients becomes lower than the rate at which they are taken by the crops from the soils. Therefore, in order to sustain a given production and productivity of foodgrains from a given area, increasing amounts of chemical fertilisers have to be applied year after year. According to some studies, the foodgrains production of about 130 million tonnes in 1980-81 in India removed about 18 million tonnes of plant nutrients from the cultivated land. Out of this total only about 10.5 million tonnes of nutrients were replenished through addition of chemical fertilisers and organic manures, thus leaving a deficit of about 7.5 million tonnes. In addition to this the Indian soils lose approximately 8.4 million tonnes of macro-nutrients annually through erosion. All this amount lost from the soil cannot be replenished in the natural course. Therefore, it is imperative that increasing amounts of nutrients should be replenished through addition of organic manures or chemical fertilisers. The scope of the former is rather limited in our country and the introduction of mechanised farming has further reduced the availability of organic manures by reducing the number of farm animals. The addition of chemical fertilisers is the only way to replenish the soil nutrients so that soils remain capable of sustaining high yields. According to FAO estimates, the consumption of chemical fertilisers in India increased from almost a zero level in 1950 to about 36 kg. per hectare in 1984 and the level presently is more than 50 kg. per hectare.

At the global scale the consumption of chemical fertilisers increased from 62 kg. per hectare in 1974-76 to 80 kg. per hectare in 1981-83. It is estimated that the future growth rate of consumption of chemical fertilisers will be about 8 per cent annually. In countries like India where the consumption is at present rather low, the rate of growth of consumption will be still higher as such populous countries must increase their food production very rapidly. However, the higher application of chemical fertilisers is not free from ecological problems. According to some of the experts, increasing agricultural productivity through increasing use of chemicals is akin to killing the goose that lays the golden eggs.

It is now-a-days a well recognised fact that not all the nutrients added to the soil in form of chemical fertilisers are used by the plants. It is estimated that only about 50 per cent of the total amount applied is used by the crops and the remaining amount is lost from the soil system without any benefit to the plants. Most of these losses are accounted for by processes such as leaching, run-off and volatilisation. The losses tend to be higher when the methods of application are faulty. Such losses not only increase the economic costs of the fertilisers, but also the chances of negative environmental impact are multiplied.

The most serious problem emanating out of the increased use of chemical fertilisers is the rapid eutrophication of surface water bodies and excessive concentration of nitrogen and other chemical compounds in water as well as in the atmosphere. Excessive deposits of nitrogen compounds may render ground and surface water resources unsuitable for human consumption. In parts of Germany and USA, groundwater has been contaminated by nitrates leaching from the upper soil layers. Excessive use of chemical fertilisers has in many cases led to the application of the law of diminishing returns. In India, though such problems have not become apparent so far, they may be encountered in future in regions of sandy soils such as the Indo-Gangetic plains, as the consumption of chemical fertilisers increases. Surface encrustations of chemicals are quite common in small patches where soil drainage is poor and the residual amounts of fertilisers accumulate on the soil surface.

Chemical fertilisers usually provide only the macro-nutrients such as nitrogen, phosphates and potash. As larger quantities of these macro-nutrients are absorbed by crops, proportionately higher quantities of micro-nutrients are also taken up from the soil reservoir. However, they are not replenished by the use of common chemical fertilisers. This leads to a deficiency of micro-nutrients in the soils leading to reduced crop yields. A study has revealed that use of 100 kg. per hectare of NPK led to depletion of zinc by 629 gms. per hectare and copper by 433 gm. per hectare. Incidentally, the zinc deficiency is the most widespread of all micro-nutrient deficiencies. Punjab and Haryana have a high level of zinc deficiency and for sustaining yields of rice, zinc has to be applied just like other chemical fertilisers in many parts of these states. Addition of zinc to the soils leads to deficiency of other micro-nutrients in the soil thus starting a whole chain of deficiencies. Use of micro-nutrients as fertilisers may well lead to toxic build ups and the Indian farmers are not educated and aware enough to monitor these applications sufficiently. The metals may enter the food chain leading to severe health complications.

Micro-nutrient deficiency affects the quality of food also. Most food samples from the green revolution states of India have been found to be deficient in elements like zinc, and recent studies conducted in Punjab indicate that the quality of wheat, maize and rice improve after zinc application leading to appreciable increase in energy values. Thus, the deficiency of micro-nutrients must be checked but the uncontrolled and unmonitored application of these nutrients cannot be allowed. Toxic build ups of these metallic substances may lead to serious environmental and health hazards.

Pesticides, Insecticides and Herbicides

Insects, pests, diseases and weeds cause large damages to crops and grains the world over, though the losses vary from one region to the other. This problem is likely to aggravate further with the increased use of irrigation which will provide congenial environment for insect pest activity. Thus the use of pesticides and insecticides will tend to increase all over the world. At present the use of insecticides in India is rather low in comparison to that in the developed countries. However, the use of these poisonous substances has increased substantially in the recent decades. In 1954-55 the amount of insecticides used in India was only 3.2 gm. per hectare which rose to 435 gm. per hectare in 1975-76. The installed capacity to produce insecticides in the country increased from 45930 tonnes in 1974-75 to 94,500 tonnes in 1983-84. The production also increased from 34,203 tonnes to 72,000 tonnes during the same period. It is estimated that the demand for pesticides in the country is increasing at a rate of at least 10 per cent annually.

Use of chemicals for pest and vector control has significantly checked plant and animal diseases and it has been an important factor in boosting agricultural productivity. The role of pest and disease control is all the more significant in tropical countries like India, where the incidence of pests and diseases is higher than in the temperate regions. Use of insecticides and pesticides has helped the country save slightly above 10 per cent of the foodgrains production. Pesticides play a major role in reducing not only the pre-harvest losses, but they also help in checking the storage losses. They are thus becoming a major weapon in the struggle against food shortage.

There are three broad types of pesticides used in the modern world; insecticides, herbicides and fungicides. The use of herbicides and fungicides is more common in the developed countries but presently rising rapidly in the developing regions. As far as the insecticides are concerned, the developing countries presently use almost the same amount as is used in the developed countries and the former are likely to outpace the later group in near future as the rate of growth of consumption is much higher in the developing countries.

The success of the pesticides, which has been quite phenomenal, in the short run, has exacted and is likely to exact an increasingly higher price in terms of side effects. Some of these side effects involve human health. Most of the pesticides used, especially in the developing countries, are broad spectrum and they are toxic to vertibrates and human populations as well. Unless adequate precautions are taken, some of them can be fatal to human, animal, fish and bird populations. Even when used most carefully and properly, they have an array of unavoidable side effects. Their persistence and ubiquity coupled with a tendency of some compounds to concentrate in organisms, as they move through food chain, may enhance their toxicity to various forms of life including man. Quite a few of the pesticides have a long life and they play a major role in reducing soil capacity to raise crops over a long run. Studies have shown that pesticides such as DDT and toxophene will remain in sandy soils intact at least 14 years after they are applied. Use of such pesticides would severely affect soil fertility if they interfere with the life of the nitrogen fixing bacteria. In fact little is known about the effect of the pesticides on these bacteria.

Some rather recent estimates by the FAO suggest that the number of unintentional acute pesticide poisoning cases in the world is about one million per year. According to another estimate, India alone accounts for about one-third of the pesticide poisoning cases reported in the developing countries. Farm workers engaged in the activity of spraying or otherwise applying pesticides are the worst affected and mostly they are exposed for long periods. Cases of deformities, cancer, blindness and diseases of liver and nervous system due to insecticide poisoning have been reported in the cotton growing areas of the peninsular region in India.

The amount of pesticides impinging on the target pests is usually a very small percentage of the amount applied, ranging from 0.1 per cent to 1.0 per cent. Thus over 99 per cent of the amount applied fails to reach the target pests and it moves into the ecosystem contaminating land, water and the atmosphere. Water contamination is a serious threat emanating out of washing down of the poisonous substances into the surface and groundwater bodies. They can also contaminate people through their ingestion of food containing concentrations at low levels. With time concentrations of poisons may take place in human bodies to serious levels through repeated consumption of contaminated food.

Some pesticides are applied to the aerial part of the plants close to harvest, especially in case of the crops of green vegetables. Large residues of these are still present in the plant parts when the crops are consumed by the people. This problem is further aggravated in a country like India where the majority of the farmers are uneducated and unaware of the consequences of mis-and overapplication of pesticides. It is common knowledge that with repeated use of a particular pesticide, the pests develop resistence to it. Under such conditions, the commonly used pesticides become ineffective and the farmer looks for some stronger pesticides to save his crop. In this effort he frequently uses a stronger dose and sometimes wrong pesticides also. There are instances when the farmers apply pesticides having a longer persistence to short duration crops. Application of pesticides meant for crops like cotton, to crops of vegetable is not rare in India. This multiplies the chances of food contamination many times often leading to severe health complications.

Comparatively new entrants in the category of poisonous chemicals used in agriculture and allied activities, such as storage of food stuffs, are the fumigation compounds like the ethylene oxide and the herbicides or the weed killers. They also contribute to contamination of environment and food in different ways. Use of ethylene oxide for fumigation of species results in presence of residues and certain of its biologically active reaction products in the foodstuffs. Consumption of contaminated bread prepared from cereals treated with fungicides has resulted in a number of instances of poisoning in different countries. Iraq experienced one such epidemic in 1971-72 when over 500 persons died in hospital due to food contamination. The total number of persons getting sick and admitted to hospitals was more than 6000. This had resulted from consumption of contaminated food and the source of contamination were the fungicides used for protecting grain from pests.

Most of the consumption of herbicides or the weed killers in the world is accounted for by the developed countries and their application has become popular in the developing countries rather recently. However, their consumption in the developing countries like India and China is fast picking up. As most of these chemicals must be added to the environment, especially to the soil, it is inevitable that some portion will certainly reach human beings via diet and other routes. Dioxin, which is the most common active agent in the herbicides, causes embryo toxicity, reproductive effects and carcinogenic effects among animals and there is definite laboratory evidence to this effect. Besides entering human bodies these chemicals may concentrate in soils retarding their plant sustaining capacity. As various types of chemicals having toxicity with respect to different types of plants accumulate, finally the soil might become too poisonous for any type of plants.

The problems emanating from applications of chemical pesticides and herbicides etc. are more common in the developing countries. Most of these chemicals having a longer persistence and affecting environment and human beings adversely have been banned or strictly restricted in most of the developed countries, like USA and many of the European countries. However, their production has not been discontinued in most of the cases and the production from these developed countries finds a ready and growing market in the developing countries. In terms of tonnage, at least 70 per cent of all pesticides used on Indian farms are banned or severely restricted in western countries and identified by the WHO as excessively toxic or hazardous. DDT and BHC are two of such pesticides which are highly toxic and not used in most of the developed countries. It is, therefore, not surprising that the highest levels of DDT and DDE in human milk and fat have been found in China, India and Mexico, in that order. Calculated daily intake of DDT and DDE by breast fed infants is also highest in the same countries, in the same order. In China this intake is much higher than the acceptable distary intake as suggested by FAO/WHO. In India also the level of this intake is at the maximum acceptable limit as per the U.N. studies and much higher in some parts according to some independent studies. A study conducted by the scientists of the Punjab Agricultural University concluded that in parts of Punjab babies were drinking 21 times the daily acceptable intake of DDT and BHC from their mothers' milk. The human tragedy at Bhopal has amply demonstrated that not only their application but even their production process can cause incalculable losses.

It has been commonly experienced that the yields of agricultural crops increase very rapidly with the introduction of pest control by chemicals and this benefit can be obtained for a number of harvests. With time, however, these gains become harder to maintain even with the use of new pesticides. This is because few pesticides are pest specific and they kill not only the target pests but also some of their natural enemies (predators). This enhances the need for further pest control by newer chemicals. In addition, an insect species which may have caused little damage because its population was kept in check by predators and parasites, can itself become a serious pest in the absence of its predators. Such resurgence of primary and secondary pests makes pest control uneconomical by repetitive use of insecticides. Resurgence of brown plant hopper in paddy due to misuse of pesticides is a common experience. Outbreak of cotton white fly due to overuse of synthetic pyrethroids against ball worms in 1985 in Malwa Plateau region played havoc with crops. According to some studies about 400 species of insects have developed resistance to one or more insecticides. Today it is feared that all pests might develop resistance to all types of chemical pesticides in due course of time. This would adversely affect the efficiency and economy of pest control the world over.

Similarly development of resistance has been observed in case of plant pathogens. As a consequence the herbicides are becoming ineffective in may cases. Up to 1965-70, FAO received very few reports of resistant plant pathogens. However, since the introduction of new synthetic fungicides, the problem increased and today as many as about one hundred plant pathogens have become resistant. Resistance has appeared even with respect to hormone pesticides in respect of which it was unthought of. Thus, in spite of the great progress made in the development of new and more effective and permanent pesticides, we cannot claim to have a pesticides to which insects or the plants will not develop resistance. If pests have to be controlled through chemical means, the farmer will have to be equipped with new pesticides at regular intervals of time. Naturally it will mean increased environmental and health hazards.

The Wayout

The foregoing analysis paints a rather bleak future for the agriculture in general and food production in particular. However, there are ways by adopting which the food problem can be solved without seriously jeopardising human health and environmental quality. If one reflects over the past a bit carefully, one would realise that most of the problems of environmental quality and human losses have resulted from either some wrong practices of landuse or from excesses indulged in by the man himself. Problems of soil erosion and land degradation are almost wholly man-made. They are a result of practices such as cultivation on steep slopes, overgrazing, leaving the land tilled and bare during the dry season and deforestation etc. Generally a lack of knowledge about land suitability and land capability is the indirect cause of many of such evils. Even today we have not appraised our land resource thoroughly in all parts of the country and whatever information is available lies in the research laboratories and the farmer who has to cultivate the land or use it for other purposes, knows nothing about the proper or improper use of this vital resource. Usually he learns from his mistakes and it is the hard way to learn. Often by the time he realises the impropriety of his actions, the damage has already been done and often irrevocably.

Our forest resources are really hard-pressed but still the productivity is on the lower side. Thus for meeting the normal requirements of the people, more than average amount of forest area or forest resources are sacrificed. This tells upon the area under forests. If the productivity of the forests can be improved and the people can be made aware of the benefits of preserving and conserving forests, the trend of declining area under forests can be checked. The greed of cultivable land has to be controlled so that there is no conversion of forest lands into cultivated lands.

Greatest harm has, perhaps, resulted from a blanket application of the imported technology in respect of mechanisation, high yielding varieties of seeds, irrigation, chemical fertilisers and pesticides etc. In many cases the cropping pattern has been altered without giving enough thought to the possible environmental consequences. One example of such a step is the introduction of rice cultivation in the semi-arid regions like Punjab and Haryana. These regions rank among the poorest parts of the country from the point of view of the groundwater resources. Perhaps, this crop can be sustained for long times in the canal irrigated areas. But it cannot be supported for many years in the well/tubewell irrigated regions of this semi-arid zone. At the same time this change in the cropping system has caused imbalance in the production of foodstuffs by removing the millet crops from the scene. A better approach would have been to develop better varieties of crops grown traditionally in such areas. They would not have caused such a high pressure on the water resources of the region and they are definitely better suited to the over all environmental conditions prevailing here. The farmers must be given to understand about the possible harmful effects of falling in for short term monetary gains resulting from such changes adopted in the cropping pattern.

Irrigation is undoubtedly a necessary input if the productivity of farming is to be improved. However, a proper stock of the situation must be taken before any blanket application of this input. Rajasthan Canal Project is today known to have forfeited at least part of its aim. If the command area cannot tolerate intensive irrigation, what is the need for a project of this gigantic magnitude. A proper appraisal of the environmental conditions such as soil drainage, must be had in advance. There should also be some system to control the use of irrigation system so that improper or misuse may be checked. If the soils of an area can withstand two waterings in a cropping season, then only such crops should be grown there which will not require more than two waterings. It will not only save the precious water resources from wastage but will also help in maintaining the quality of environment.

Chemical fertilisers and chemical pesticides have been considered the most prominent factor causing environmental contamination. Undoubtedly they are. The problem in this regard is; can we do without them? The answer is an obvious No. The only solution to the problems emanating out of the use of chemical fertilisers and pesticides lies in the methods of their application and a judicious use of these inputs. The amount of fertilisers used is not likely to decrease, but the proportions actually used by the plants can definitely be increased by changing the methods of application. Some studies in India have shown that the proportion of the fertilisers being used by the plants, can be increased if smaller quantities of fertilisers are applied more than once, instead of applying the whole amount in one go. This also reduces the amount of fertilisers lost from the soil system thereby checking pollution of environment and reducing health hazards at the same time. This will enhance the efficiency of the fertilisers and bring down the costs involved. The farmers should be provided adequate information and guidance concerning the type of fertilisers that should be used, the required amounts thereof and also the proper quantities and time interval between successive applications. The use of fertilisers should be monitored, if possible. Care should also be taken to provide micro-nutrient either as a constituent of fertilisers or as independent sources.

Chemical pesticides should be used with utmost care and control. Unfortunately the situation in this regard is most depressing. A farmer can buy any type of pesticide, in any quantity and he can apply it to any crop, in any amount and at any time. There is no mechanism to check this menace. The producers of pesticides are also literally free to produce chemicals of any toxicity provided they print a warning on the pack. Even under controlled use, these substances have been causing problems in the developed countries, and without any control, as is the situation in India, they can cause untold problems and health hazards. Like fertilisers, the pesticides can also be used in a smaller doses by such methods as can make them more lethal to the target pests and the residues can thus be checked. There is a great need to educate people about not only the methods of application but also about the choice of pesticides for various purposes and different crops. There is definitely a need for strict monitoring at the production stage as well as at the application stage. The pesticides having a long persistent life, such as DDT, DDE and BHC must not be freely available. If not banned, they should be strictly controlled. A farmer cannot be convinced easily to not to use a cheap and strong pesticide if it is available freely in the market. Thus the only wayout is the legal steps controlling their availability and applications.

There is also a need for monitoring the efficiency of pesticides over time to watch out for the resistance that the pests and the plant pathogens might develop. There should be some mechanism to advise farmers in the changes in the chemicals to be used. There are also ways in which the consumption of the chemical pesticides may be checked. Biological pest control is one such method and it should be adopted wherever feasible. Traditionally, the Indian farmers have adopted various techniques to save their crops from pests. One such method is to grow two crops simultaneously. The crops are such as are susceptible to a common pest, but to varying degrees. One example of a pair of such crops is cotton and lady finger. In parts of Uttar Pradesh, the farmers occasionally grow both these crops simultaneously. Being more susceptible to common insects, ladyfinger plants attract all the insects and when the insects have concentrated on these plants, they are uprooted from the field and burnt. This saves the cotton crop at a very low cost and in fact no environmental cost. Such practices can be encouraged so that some of the pests can be controlled without using any chemicals. In any case, the pesticides should have a short active life and as far as possible they should be bio-degradable.

Finally, one must not forget that the food problem equation has two sides like any other equation; a demand side and a supply side. All methods of increasing food production aim at balancing this equation by augmenting the supply side. The demand side of the equation is represented by the food requirements and this side is controlled by the size of population. If the demand side is allowed to grow unchecked, no efforts on the supply side can find a solution to balance this equation. Therefore, the more appropriate approach to this equation would be to attack the problem from both sides. On the one hand we should try to increase the agricultural production and on the other the efforts should be made to control the growth of population. No technology can keep on feeding an ever-increasing population on the production from a fixed area of land. Thus, the efforts on the front of population control must continue simultaneously.

Lastly, but not the least, before any innovation is adopted, all possible environmental and human consequences of it must be carefully evaluated and we should be prepared to abandon or not to adopt any technology if it is detrimental to environment or poses any threat to human lives.

REFERENCES

Bandyopadhyay, J., et. al., 1985 : *India's Environment-Crisis and Response;* Natraj Publications, Dehra Dun.

Centre for Science and Information, New Delhi : *The State of India's Environment* 1984-85, Second Citizens' Report.

El Himavi and Hashmi, M. H., 1987 : *The State of the Environment* (UN Environment Programme), Butterworths, London.

Goswami, N. N., 1988 : 'Optimising Soil Environment for Sustained Agricultural Production' in Khoshoo, T.N. (Ed.), *Perspectives in Environmental Management,* New Delhi.

Jain, H., 1988 : *Contribution of New Technology in Agriculture*, Kurukshetra, October.

Khoshoo, T. N., (Ed.), 1988 : *Perspectives in Environmental Management*, New Delhi.

Mehta, B., 1988 : *Indiscriminate Use of Technology Can be Counter Productive*, Kurukshetra, October.

Misra, U. S., 1989 : "*Insecticides and Environment: An Observation*" Kurukshetra, January.

Owen, O. S., 1985 : *Natural Resource Conservation,* MacMillan Publishing Co. New York.

Singh, K. P. Groundwater and Environment in R. K. Sapru (Ed.), 1987 : *Environmental Management in India*; Ashish Publishing House, New Delhi.

possible environmental and human consequences of it must be carefully evaluated and we should be prepared to abandon or suitably modify any technology if it is detrimental to environment or poses any threat to human lives.

REFERENCES

Bandyopadhyay, J. et al. 1985. India's Environment: Crises and Responses. [illegible] Publications, Dehra Dun.

Centre for Science and Environment, New Delhi. The State of India's Environment 1984-85 Second Citizens' Report.

El-Hinnawi and Hashmi, M. H. 1987. The State of the Environment (UN Environment Programme), Butterworths, London.

[illegible] 'Opening Soil Ecosystems for Sustained Agriculture' [illegible] UNESCO.

[illegible]

[illegible]

[illegible]

[illegible]

15

AGRICULTURAL GROWTH AND RURAL POVERTY

S. F. KORISHETTAR

THE PRESENT paper deals with the shift in official planning policies, and the government's renewed rural development efforts in recent years. It examines the inter-relationship between agricultural growth and rural poverty and unemployment, and analyses the role of irrigation and ancillary industry in the process of economic development.

The results of planning during the first phase of development (1951-1974), reviewed in several studies show that material benefits accumulated in a few major urban enclaves of economic growth but failed to "trickle-down" to the poor regions. And the adoption in the mid-1960s of the New Agricultural Strategy (HYVs Programme) also resulted in uneven growth, by favouring better-endowed regions and failing to alleviate rural poverty and unemployment in other regions. But the last two years of the Fourth Five Year Plan (1973-74) seemed to signal a new state of affairs. The urgency of the situation at that time called for heightened political commitment to a rurally-based strategy of national development.

A search started for new and substantially different approaches to national development. It focused on the problems of the rural people who make up the majority of the population and whose needs had been largely ignored by the authorities for the almost two and a half decades of planned development in India (Misra and Nataraj, 1981). The new commitment confronted Indian policy-makers with many practical problems; not least, how to engage the millions of rural households in the development process. As Friedmann (1974) has pointed out: "It cannot be denied that the task of developing the rural economy is a task of enormous complexity". He further notes: "How does one reach the hundreds of thousands and even millions of individual production units that are scattered over the entire territory of a nation? The problem is difficult enough in rich countries. It is infinitely more so in countries that are poor" (Friedmann, 1974).

In the context of the proposed new strategy of rural development, India's Sixth Five Year Plan (1980-85) makes an important observation. "It is a relatively simple matter to prepare projects for setting up industries or power stations or expanding capacities of ports or airlines or telecommunication systems. A limited number of experts, planners, contractors, and workers are needed to implement such schemes. A programme of rural development, on the other hand, involves investment decisions by hundreds and thousands of individuals and their ability to evoke response of millions of potential beneficiaries. It calls for the active participation of a large number of people, a substantial proportion (65%) of whom are illiterate and unskilled. In such a situation, the government machinery has to be much more active than before.

As indicated by Krishna (1986) successful rural development in poor countries is now recognised to involve a major re-ordering of investment priorities and adoption of a new spatial framework for the formulation and execution of the appropriate policies. The Fifth (1974-79) and Sixth (1980-85) Five Year Plans declare their objectives to include reducing geographical imbalances between the regions, income inequalities, and rural poverty and unemployment through direct investment in rural areas. The Seventh Five Year Plan (1986-1990), proclaims an intent to benefit not only primarily the small and low-income farmers but also those living in agriculturally based service towns, and to commit industrial development to a long-range strategy for balanced agricultural and industrial development (Iqbal and Mohammad, 1986). These objectives appear to be similar to those approaches now proposed by regional theorists, such as Stohr and Taylor (1981) and Friedmann and Douglass (1975).

The "20-Point Programme" declared in 1975 (revised in 1980) has been one of the principal guiding factors in the formulation of the recent Five Year Plans. The National Programme of Minimum Needs (NPMN) was incorporated in the Fifth Five Year Plan. It envisages a frontal attack on the problem of rural poverty and unemployment by attempting to allocate adequate resources for the creation of diversified and dispersed employment, income generating activities and direct provision of minimum needs in the rural and backward areas (Planning Commission, 1974-79). It was declared that initial attention should be concentrated on elementary education, public health facilities (which include preventive medicine, nutrition, clean drinking water, provision of developed home sites for the poor and landless labourers in rural areas and slum improvement) and rural-to-rural all-weather roads and rural electrification. It was anticipated that implementation of such a programme would go a long way towards establishing a healthy environment for social well-being and economic growth and also create a feeling of participation in national development among the rural masses.

The thrust of the revised "20-Point Programme" included in the Sixth Five Year Plan not only continues to be on providing better living conditions for the poor and less privileged sections of the population but also aims at an all-round improvement in productivity and the growth of national economy. It lays particular stress on the pace of agricultural development, taking steps to: remove regional and cropping imbalances, evolve new varieties of seeds, expand irrigation facilities, foster decentralisation and the organisation of consumer goods industries, village and small-scale industries and the provision of basic amenities — health, education, house sites and so on. All such measures are considered to go a long-way towards improving the economic and social conditions of the rural masses, and are also intended to increase rural employment, strengthen the resource base of the rural poor and raise their incomes.

Many studies have demonstrated that approaches which aim to provide basic infrastructure for agricultural development and minimum needs and services tend to have a better than average performance in terms of economic growth (Ahluwalia and Chenery, 1974; Chenery, 1980; Hicks and Pstreeten, 1979, and Hicks, 1980). No doubt the main problems that the rural sector of the economy has been confronted with can be successfully tackled by carrying out such far-reaching reforms. But, the extent to which programmes embodying this new shift in official policy towards national development can effectively redress the contradictions created by continued reliance on a strategy of accelerated industrialisation, needs to be examined empirically at the village level.

Agriculture and Equitable Growth

The substantial literature which has been produced in recent years provides two broad perspectives on equitable growth. These concern the overall relationship between development and distribution, and the role played by the agricultural sector in fostering equitable growth. However, there is controversy about whether improvements in agricultural technology that expand food output improve or worsen the lot of the rural poor. It is argued that the poor-to-marginal farmers, who have little land, produce food largely for themselves, and the landless labourers, employed in agriculture and other construction services, depend on market purchases for the major share of their consumption needs. Under these circumstances, increased agricultural production and higher producer prices will benefit the rich farmers.

However, we argue that the negative effects of increased agricultural production will be overwhelmed by its positive effects. Basically, the question which needs to be considered is: what causes the incidence of rural poverty to increase or decrease? No single criterion can define precisely what poverty is. There are problems with the measurement of poverty, whether its incidence be determined by per capita income, minimum needs, consumption levels, or by access to health, education or land and so on. To identify the real causes and nature of poverty in the Indian context, and the means to alleviate it, needs village-based empirical research that may reveal its dynamic nature. Some systematic efforts have been made in this direction (Srinivasan, 1974; Mellor and Desai, 1986; Ahluwalia, 1977; Narain, 1961, and Narain, 1965). They all deal with the conceptual basis of the relation between agricultural production and poverty. Most of the studies conclude that the condition of rural poverty is the direct result of farm output. One empirical study has observed that the fluctuations in the Indian poverty rate could be explained by the growth of agricultural production and changes in the prices of agricultural commodities consumed by the rural poor (Narain, 1965). Another study, examining India's National Sample Survey (NSS) consumption expenditure data, has found a statistically significant relationship between growth in agricultural production per capita and reductions in rural poverty (Narain, 1965).

The Indian Planning Commission (1977) prepared estimates of the population below the poverty line for the period 1972 to 1978, for a poverty line defined in terms of the consumption levels which would allow a daily minimum per capita intake of 2,400 calories in rural areas and 2,100 calories in urban areas. Using this poverty line and NSS consumer expenditure data, the Planning Commission observed that the percentage of the population below the poverty line declined from 52 per cent in 1972 to 48.8 per cent in 1978 for rural areas. The estimated figures of the Planning Commission reveal that the poverty rate is indeed a zigzag one. It has risen and fallen several times during the past 30 years (Table 15.1). The percentage of the population below the poverty line declined sharply from 53.4 per cent in 1957-58 to 42.3 per cent in the early 1960s, but then climbed sharply to 58 per cent in 1967-68. It then declined again, reaching 48 per cent in the mid-1970s. The Government's most recent estimates indicate that about 40 per cent of the Indian population live below a minimal poverty line.

TABLE 15.1: **Rural population below the poverty line**

Years	Rural Population below poverty line (in %)
1957-58	53.4
1959-60	48.7
1960-61	42.0
1961-62	42.3
1962-63	49.1
1963-64	48.9
1964-65	50.4
1965-66	51.1
1966-67	57.4
1967-68	57.9
1968-69	53.5
1969-70	52.3
1970-71	49.1
1773-74	47.6
1977-78*	48.9
1982-83**	40.1

* Government of India, Planning Commission, 1977, New Delhi.
** *The Economist*, News weekly, Aug. 13, 1987.
Source: Journal of Development Studies, Vol. 13, July 1977.

If most people agree that the poverty zigzag is a function of agricultural output, it seems logical to argue that rural poverty and unemployment are largely determined by Indian weather conditions. The basis for our argument is that about 73 per cent of the cropped area is still unprotected by artificial irrigation. Therefore, rainfall is the predominant variable governing agricultural output. For example, if we look at foodgrain production over a period of 30 years, the large fluctuations in output (Table 15.2) are largely the function of favourable weather conditions. Those years of increased agricultural production are the years of timely rainfall. In other words, the principal factor behind the increased agricultural output is not the increased area under food crops, nor the increased use of superior technology, but the benevolent "Rain-God" — monsoon.

TABLE 15.2: Foodgrain production in India (1950-1982)

(in million tons)

Year	Rice	Wheat	Other Cereals	Pulses	Total Foodgrains
1949-50	25.1	6.8	18.9	10.0	60.8
1950-51	22.1	6.8	16.9	9.2	50.0
1951-52	22.6	6.3	17.5	9.1	55.6
1952-53	24.3	7.6	20.0	9.8	61.8
1953-54	29.8	8.1	23.3	11.1	72.3
1954-55	26.5	9.1	23.4	11.6	70.7
1955-56	28.7	8.9	20.1	11.7	69.3
1956-57	30.2	9.5	20.6	12.2	72.5
1957-58	26.5	8.0	22.0	10.1	66.6
1958-59	32.0	9.9	23.6	13.2	78.8
1959-60	31.8	10.3	23.1	11.9	77.1
1960-61	34.6	11.0	24.0	12.7	82.3
1961-62	35.6	12.1	23.1	11.7	82.4
1962-63	33.4	10.8	24.6	11.5	80.3
1963-64	37.0	9.8	23.8	10.1	80.7
1964-65	39.3	12.3	25.4	12.4	89.4
1965-66	30.6	10.4	21.4	9.9	76.3
1966-67	30.4	11.4	24.1	8.3	74.2
1967-68	37.6	16.5	28.8	12.1	95.1
1968-69	39.8	18.7	25.2	10.4	94.0
1969-70	40.4	20.1	27.3	11.7	99.5
1970-71	42.2	23.8	30.5	11.8	108.4
1971-72	43.1	26.4	24.6	11.1	105.2
1972-73	39.2	24.7	23.1	9.9	97.0
1973-74	44.1	21.8	28.8	10.0	104.7
1974-75	39.6	24.1	26.1	10.0	99.8
1975-76	48.7	28.8	30.4	13.0	121.0
1976-77	41.9	29.0	28.9	11.4	111.2
1977-78	52.7	31.7	30.0	12.0	126.4
1978-79	53.8	35.5	30.4	12.2	131.9
1979-80	42.3	31.8	27.0	8.6	109.7
1980-81	53.2	36.5	29.0	11.2	129.9
1981-82*	53.3	37.5	31.1	11.5	133.4
1982-83*	47.1	42.8	27.7	11.9	129.5
1983-84*	59.8	45.1	33.9	12.7	151.5

* Figures for 1982-1984 are from *Centre for Monitoring Indian Economy* (CMIE) (1985), Agricultural Production in India. State-wise and Crop-wise Data: Bombay.

Source: Government of India, *Bulletin on Food Statistics 1982*, Ministry of Agriculture, New Delhi.

Secondly, there are marked spatial variations in foodgrain production. A small number of states in India contribute the major production of foodgrains. Only five states — Uttar Pradesh, Punjab, Haryana, Tamil Nadu and Madhya Pradesh — with 46.7 per cent of the total area under foodgrain production, contribute an annual average of about 52.3 per cent. And again there are great sub-regional disparities within these states. A very high degree of concentration of output is found only in a few favoured districts. The spatial distribution of area according to levels of productivity in foodgrains output is presented in Table 15.3 and Fig. 15.1. It is clear that the high productivity districts are mainly concentrated in the states of Punjab, Haryana, Tamil Nadu, some parts in Karnataka, and Kerala.

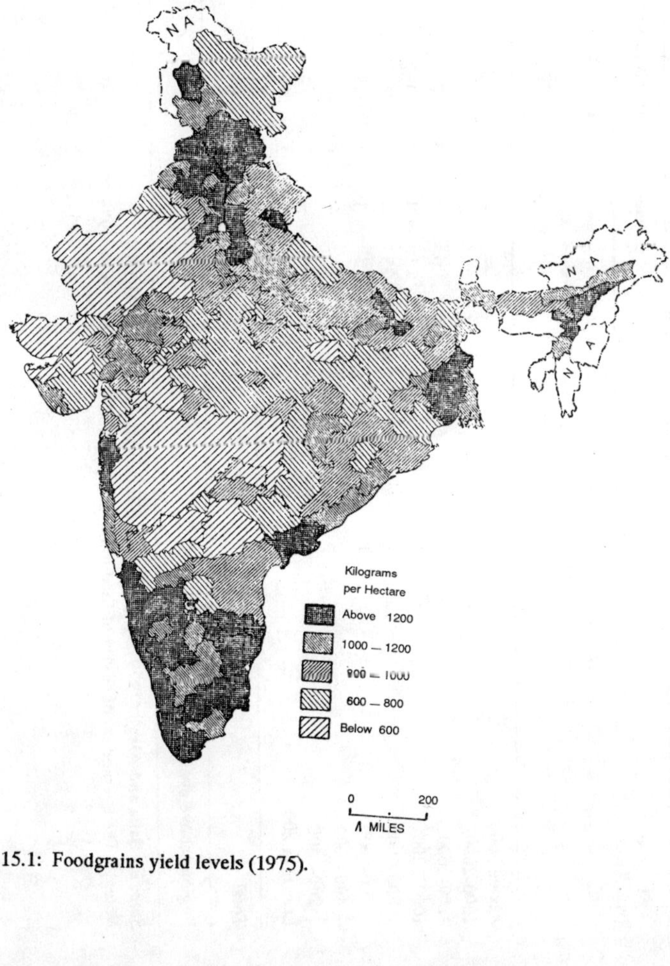

FIG. 15.1: Foodgrains yield levels (1975).

TABLE 15.3: Spatial distribution of areas according to levels of productivity in foodgrain (1978-79)

Yield Levels (Kgs/Hect)	Number of Districts in Various States																	
	Punjab	Tamil Nadu	Har-yana	Andhra Pradesh	Kerala	Karna-taka	West Bengal	Uttar Pradesh	Jammu & Kashmir	Maha-rashtra	Assam	Bihar	Hima-chal Pradesh	Gujarat	Rajas-than	Madhya Pradesh	Orissa	Total
Above 2000	5	2	—	—	—	—	—	—	—	—	—	—	—	—	—	—	—	7
1600-2000	4	5	1	1	1	2	1	—	—	—	—	—	—	—	—	—	—	15
1200-1600	2	1	2	3	6	8	6	6	1	3	2	2	1	—	—	—	—	43
1000-1200	—	4	2	1	—	3	8	11	1	1	3	2	—	4	3	1	1	45
800-1000	—	—	1	4	—	1	—	25	—	1	3	6	—	3	8	11	5	68
600- 800	—	—	1	3	—	3	—	7	1	1	—	5	—	8	3	27	5	64
400- 600	—	—	—	5	—	2	—	—	—	8	—	—	—	2	5	4	—	26
200- 400	—	—	—	—	—	—	—	—	—	11	—	—	—	1	4	—	—	16
Less than 200	—	—	—	—	—	—	—	—	—	1	—	—	—	—	3	1	—	5
Total	11	12	7	17	7	19	15	49	3	26	8	15	1	18	26	44	11	289
Regions above average	11	12	5	6	7	13	15	24	2	5	6	6	1	5	5	4	2	229

Source: Balla and Alagh (1979), *Performance of Indian Agriculture*, Table 13(b).
Note: The average productivity of foodgrains in India is 915 Kgs/Hect., as on 1978-79.

The JNU-PPD (Jawaharlal Nehru University's Perspective Planning Division) study revealed the vast differences in yields across the country. Of 282 districts studied, 69 accounted for 20.5 per cent of the cultivated area in 1973 but contributed 38.7 per cent of the national foodgrain output. At the other extreme, 83 districts accounted for 33.8 per cent of the total cropped area but produced only 14.6 per cent total output (Dantwala, 1986). Thus, high, medium and low output areas could be identified throughout the country. It has not been possible for us to analyse fully in this report the various factors that are responsible for high levels of growth in certain areas and lack of growth elsewhere. Nevertheless, some patterns emerging from the JNUPPD study reveal that the areas with high agricultural productivity are significantly associated with areas of assured water supply. The critical importance of developing water works to achieve sustained growth in Indian agriculture, therefore, cannot be denied. The farmers and agricultural labourers are at the mercy of the monsoon. If precipitation fails to occur in time the whole economy collapses. The prices of foodgrains, pulses, edible oils soar to new heights. The poor farmers have to go out of their villages to hire out their labour for wages in order to meet their livelihood. This happens because the "Kharif" output — the monsoon harvest (June-October) accounts for about 60 per cent of the annual production of foodgrains and about 55 per cent of oil seeds production. Therefore, the development of irrigation works is an essential prerequisite to creating regional equity and employment within the rural sector.

Land-Reforms and Equitable Growth

There is no example or model which Indian planners can exactly follow in solving the problems of reconciling the development of the rural areas with the growth of the national economy. Nor is there an agreed theory relating rural development and the fair distribution of incomes in individual households. There are, of course, some developing countries which have grown relatively rapidly and avoided an absolute worsening of rural poverty and unemployment. The People's Republic of China, South Korea, and Taiwan are cited most frequently as cases of equitable growth. However, the conditions under which the agricultural sector has promoted more equitable growth in these countries has to be understood. All three began their phases of modern economic growth with either revolution or war. In China, the

Communist Revolution in 1949 was followed by the radical re-distribution of endowments through land reforms and nationalisation (Gurley, 1975). In South Korea, the Korean War was followed by a significant land reform (Westley, 1986). In the case of Taiwan, total disruption of the Taiwanese social order by the Nationalist Chinese fleeing from the mainland was also followed by a thoroughgoing land reform. Thus, radical changes represented by major redistributive reforms seem to have been the prerequisite to agricultural growth and the reduction of rural poverty, unemployment and inequality in these countries. Another characteristic of these three countries is their unified cultural heritage, which has a considerable impact on economic performance. India is a country of subcontinental size, with great geographical, ecological, cultural, religious, political and economic diversity. Under the circumstances, whether the Korean or Chinese model of rural development is applicable to the prevailing conditions of India has to be analysed.

However, there are contrasting views about such policy measures. One group of economists argues that basic structural changes through redistributive land reform is the key element of any strategy for equitable growth (Gurley, 1975; Lardy, 1980; Mason, 1980; Fei, et al., 1979; Perkins, 1980, and Griffin, 1979). According to them, agricultural development can contribute substantially to equitable growth only through a structural shift in land endowments.

The opposing group maintains that equitable growth is possible without radical redistributive land reforms (Lipton, 1977; Mellor, 1978; Johnstone, 1982, and Mellor, 1976). Holding a similar view, we argue here that radical land reforms may not do much to increase agricultural production or ameliorate rural poverty and unemployment in the case of India.

Land reform has been suggested by Indian planners, and to a certain extent adopted, as one of the measures to reduce rural poverty and unemployment. Two principal types of land reform have been adopted: (1) imposition of a ceiling on land-holdings and redistribution of surplus land to the landless, (2) tenancy reform. In the context of ceilings it has to be understood that they have primarily a redistributive impact. That is, imposition of the ceilings takes away land from some, the relatively rich, and hands it over to others, the relatively poor. It does not create any new assets, nor does it create new employment — at least immediately. Whether it increases total employment in the long-run is also the question to be analysed. In addition to this, in India the poor are so many and the rich are so few that even if at the very low ceiling limit, the land of the rich were transferred to the poor, no substantial benefit is likely to occur. In fact there is a view that such redistribution of land will hamper the growth of agricultural production. This measure may still be considered desirable, as an egalitarian strategy, but that is a separate issue.

Secondly, reform of the tenancy system also has limited relevance because the tenancy picture in India is quite distinctive. Purely landless tenants are relatively few in number. Their proportion among the tenant class is not more than one-fifth. About 70 per cent, or an overwhelming majority of the tenants, are themselves land-owners, who take some land on lease. And again, among landlords, most are owner-cultivators who lease out some part of their land. Pure landlords are also very few in proportion. The majority of the people benefited by the enforcement of tenancy laws are medium size land-owners who took some land on lease from the poor.

Some regional analysts, therefore, suggest that increased agricultural production and poverty reduction is feasible without land reform if countries shift away from the capital intensive "Industry-First" policies of the past, under which agriculture was neglected or actively discriminated against (Johnston and Clark, 1982). Johnston distinguishes two basic agriculture strategy options, "bimodal" (or dualistic) and "unimodal". According to him, the collectivisation pattern of agriculture in the Soviet Union was clearly bimodal, whereas collectivised agriculture in China has evolved along essentially unimodal lines, based on the decentralisation of decision making to relatively small production units and the use of labour-intensive techniques (Johnston and Clark, 1982).

The distinction between the bimodal and unimodal patterns of development is based principally on the size distribution of farming units and the technologies adopted. A bimodal pattern, involving excessively large state and collective farms, results in a slower rate of growth of opportunities for productive employment within the agricultural sector because of its labour-saving technology. Under the "unimodal" pattern of development Japan, Taiwan, and South Korea were able to achieve widespread increases in productivity, employment and incomes within the existing framework of the small-scale farm. The "Japanese model" of agricultural growth is based on continued use of labour-intensive, technologies, innovations in HYV's seeds, and fertiliser-responsive crop varieties which are figured in the green revolution (Johnston and Clark, 1982). Johnston's contention is that equitable growth is possible with an adequate "sectoral shift" toward agriculture and allied industries. Lipton (1977) contends that the "better strategy is to concentrate first upon high-yielding mass-rural development supported by such selected ancillary industry as rural development makes viable" (Lipton, 1977).

In the existing Indian situation, even an effective land reform programme would not be enough. Simultaneous programmes to increase agricultural productivity, to expand irrigation and the area under HYVs and to develop the basic infrastructure are needed for effective rural development. Even with increased agricultural production and extended areas under HYVs, the agricultural sector may not be able to absorb the growing labour force in India. For this reason, it is important to foster agriculturally related activities in rural areas, for example, poultry raising, dairying, sericulture, fisheries, agro-processing industries, and consumer goods industries. Employment opportunities could also be created by building roads, minor irrigation tanks, wells, canals, soil conservation bunds, terraces, warehouses for storing foodgrains and in extensive programmes of tree planting and pasture development.

Greater emphasis on afforestation and pasture development is necessary to reduce soil losses and maximise groundwater recharge. These works can be considered investments rather than expenditures because they promise to increase future yields. They are also highly labour-intensive, complementary to agricultural productivity and require few resources. Rural poverty in India is associated with unemployment. Sustained expansion of activities such as these would stimulate the growth of employment throughout the rural economy. Improved agricultural productivity would release resources to produce goods in ancillary rural industries and increased agricultural income would, in turn, increase the demand for such goods and services. Indian planners now turning to the creation of such work opportunities — what are called "Public Rural Works". Such programmes had been occasionally undertaken in the initial plan periods, but they were sporadic and the financial outlays on them very small.

Irrigation as a Source of Equity

Another basic question facing Indian planners is: can the "Green Revolution" contribute to reducing the inter-regional disparities in income and creating employment within agriculture? It is argued here that if investment in irrigation and dryfarming technology is directed properly, if priorities are re-ordered from large-scale, costly projects to small-scale irrigation schemes, and if concerted efforts are made to reduce the cost of HYV technology (which is economically appropriate to an average Indian farmer), it is possible to create greater equity and more employment opportunities within the rural sector.

First of all, the ultimate irrigation potential of available Indian river water, surface and groundwater should be harnessed and fully utilised. Second, and most important, vulnerability to drought should be reduced through suitable changes in agricultural operations in rain-fed and drought prone regions. The development of technology for conserving soil moisture, and the evolution of new varieties of seeds and fertilisers for semi-arid and rain-fed areas are critical. It is in this field that research and extension services need to be concentrated. Fortunately, the endeavours of the International Crop Research Institute for the Semi Arid Tropics (ICRISAT) are beginning to yield promising results. The Director General of ICRISAT suggested in New Delhi in 1981 that: ". . . It is not difficult to believe that the low yields (below 800 kg/ha in rainfed areas in the case of India) can be increased by 60 to 100 per cent". He further said: "Pilot scale testing is being done in farmers' fields by ICRISAT and the results are highly promising" (Swindale, 1981). It appears that the agricultural picture is potentially bright in the rain-fed areas. With improved land management and adoption of new dry-farming technology, it would be quite possible for an average farmer to increase gross returns by Rs. 3,086 per hectare (Swindale, 1981). Therefore, large tracts of dry-land could be major contributors to growth to increasing employment, and to the reduction of rural poverty.

Similarly, there are bright agriculture prospects within the controlled water supply areas. Most of the growth in agricultural production over the past two decades has resulted from increased productivity of cropland. This has resulted from the use of a wide range of physical, biological and chemical inputs — improved farm implements, fertilisers, pesticides, the breeding of high-yield varieties, and the development of fast-growing cultivators that allow double and even triple-cropping. Based on achievements in demonstration plots, yields of most crops could be increased 4 to 11 times (Krishna, 1986). However, behind this high-input agricultural system, irrigation plays an important role. The increased productivity of existing land is attainable if the remaining large percentage of cropped land is brought under irrigation. Most studies have shown that irrigation is an important input and also a prerequisite for full adoption and extension of the HYV programme. The importance of irrigation is shown in Table 15.4, in which one *Gramsevak* (Village Level Worker — VLW) has calculated the yields of rice per acre for different combinations of the three inputs of irrigation, chemical fertilisers, and HYV seeds, in a village in the State of Karnataka.

TABLE 15.4: Response of rice crop to inputs — irrigation, fertiliser and HYV seeds (as on November 1984)

Irrigation	Chemical Fertiliser	HYV Seeds	Total Yield Per Acre (in Quintals)	Per cent Increase in Yield
0	0	0	4.8	—
0	0	1	—	—
0	1	0	5.4	12.5
0*	1	1	—	—
1**	0	0	6.0	25.0
1	0	1	—	—
1	1	0	9.0	87.5
1	1	1	10.5	118.7

Source: Data derived from local observations made for two successive 'Kharif' seasons by VLW of Block Development Office, Haveri, Karnataka State, India.

Note: * Input not applied.
** Input applied.

The data shows that irrigation alone significantly raises yields (about 25 per cent in this village), which points to the importance of irrigation as a prerequisite for the adoption of other inputs. The table shows that without irrigation, fertiliser only increases yields by about 12 per cent. With irrigation, on the other hand, fertiliser increases yields by 50 per cent. The empirical data calculated at the micro level is in consistent with findings elsewhere at macro levels. It is almost universally stated in the development literature that the same relation exists (Mellor, 1976). And one notices in the Indian countryside that without irrigation, villagers are afraid to invest in modern inputs, and barely adopt the HYV programme. Chemical fertilisers and pesticides are quite harmful to the crops if the soil contains insufficient moisture. If the farmers are questioned about why they do not adopt new varieties of inputs, many of them will explicitly mention the need for timely water supply.

The fact is, in Indian agricultural operations, timely water supply or precipitation is necessary, particularly in a one-week period at the beginning of the "Kharif" season (first week of June). Crops need an adequate start to resist the wind, heavy rainfall (monsoon) or high temperature that may follow thereafter. And it is necessary to get crops started in time to permit harvesting before the next set of crops (the "Rabi" season — winter harvest) are to be sown. Delays in sowing cause severe crop damage either by diseases or by retreating monsoons (winter rain). Thus if the monsoons fail to occur during the precise period needed, farmers are subject to adverse consequences in the following period. Timely water supply is necessary to avoid this risk period.

Thus, lack of irrigation is a critical factor in preventing farmers from achieving increased gross returns per acre. This being the so, the Indian economic planners have been severely deficient in supplying adequate resources for the extension of irrigation in large tracts of rural areas. As every farmer in the villages know, those who cultivate land with better access to irrigation get higher returns at the same input level than those who cultivate land without irrigation. This simple principle appears never to have been imbibed by the urban-based spatial planners during the initial four Five Year Plans.

If we have a look in Punjab, Haryana, and in some parts of Rajasthan and Tamil Nadu, irrigation has made the largest contribution to increased agricultural production. In these areas, it can double or triple yields during the main growing season, make even a third crop possible and can sharply reduce the risk of crop failure due to untimely rainfall. Irrigation has raised the productive capacity of the land in these areas and several studies have shown that labour absorption and equity have also been increased on Punjab and Haryana regions (Westley, 1986; Singh and Shidhu, 1976).

The most important downstream effects of irrigation are observed by the World Bank study team led by Turnham et. al. (1982). Empirical research conducted in the Muda region of Malaysia shows what additional investments and employment may be induced from the irrigation. It is calculated that farmers' income rose by about 73.4 per cent and landless farm workers' earnings rose by 70.1 per cent. Farmers' increased income from paddy generated demand for goods and services equivalent to 43.2 per cent of the total benefits of the irrigation scheme. This demand was in such sectors as, housing and other construction works, rice-milling, commerce, road transport, and hotel and restaurant services. All such activities are labour-intensive.

Higher earnings in these activities in turn multiplied jobs and incomes for workers in still other parts of the economy. Thus the direct benefits of investment in irrigation are not hard to identify within the rural economy. By the increased foodgrain output, essentially, the poor consumers will benefit from lower food prices.

Impact of Green Revolution on Employment

It is commonly observed that agriculture in the developing countries is faced with the problem of disguised unemployment. This is because of the seasonal nature of the traditional agriculture and the low level of intensity of cropping. For instance, there is a certain season in India (April to mid June) when there will be practically no agricultural work for the labourers. The solution to this problem lies in utilising the surplus labour force more productively to build up long-term assets within the rural sector. This may, however, be a partial solution to the unemployment problems in rural areas. The problem of seasonal unemployment could be solved by undertaking measures to provide increased and continuous employment on the farm. This can be done by the adoption of modern farm technology or widespread introduction of "green revolution".

A number of empirical studies undertaken in Asia with special reference to India have noted that labour use in the cultivation of HYVs was considerably higher when compared to traditional varieties. (Herdt, 1980; Bartsch, 1977; Chinnappa, 1977; Shidhu, 1974 and Lavania et. al., 1977). Adoption of HYV crops resulted in higher labour inputs per unit of cropped area and more continuous employment for the labour on the farm throughout the year. Chinnappa's (1977) study relating to HYVs of rice in North Arcot district of Tamil Nadu, observed that the increase in demand for hired labour from cultivation of HYVs, as compared to traditional varieties of rice, was 22 per cent. In absolute terms, every acre that went over to HYVs of paddy generated an additional demand for 17 per cent person/days of labour (Chinnappa, 1977).

Lavania et. al. (1977) estimated the impact of the green revolution on crop-specific employment. The study covered a stratified random sample of 120 farmers, of whom 60 were adopter farmers of new farm technology and 60 were non-adopters. The farmers were selected from 10 villages in Madhya Pradesh. The comparison of rates of growth of employment on adopter and non-adopter farmers is analysed in their study. The four major crops — paddy, corn, wheat and sorghum (jower) — were selected for study. It can be observed from the Table 15.5 that employment for all the four crops is higher on adopter farms. For the farm business as a whole, adopter farms provide 18.7 per cent more employment in terms of man/days per hectare. The cropwise rates of increased employment are: paddy (13.59%), wheat (21.00%), sorghum (21.57%) and corn (29.14%). The study also indicates the significant variations between adopter and non-adopter farms with regard to the seasonal rates of increased farm employment.

TABLE 15.5: **Employment of human labour per hectare in farm business and major crops by size of farm**

Farm size (hectares)	Adopters	Non-adopters	Adl. utilisation on adopter's farm	Percentage increase over non-adopters
Farm Business :				
Below 2.00	124.89	107.12	17.77	16.59
2.00 –400	121.54	102.92	18.62	18.09
4.00 & above	111.55	93.59	17.96	19.19
All Farms	113.47	95.54	17.93	18.77
Sorgum (Jowar)				
Below 2.00	97.68	82.47	15.21	18.44
2.00 – 4.00	88.37	73.53	14.85	20.18
4.00 & above	80.93	65.85	15.08	22.90
All Farms	82.05	67.49	14.57	21.57
Paddy :				
Below 2.00	116.66	98.83	17.72	17.93
2.00 – 4.00	103.45	90.63	12.82	14.14
4.00 & above	92.27	81.45	10.82	13.28
All Farms	96.79	85.21	11.59	13.59
Corn (Maize) :				
Below 2.00	119.42	95.24	24.18	25.39
2.00 – 4.00	114.63	86.94	27.69	31.85
4.00 & above	105.35	81.17	24.18	29.79
All Farms	107.61	83.33	24.28	29.14
Wheat :				
Below 2.00	173.57	148.75	24.82	16.69
2.00 – 4.00	162.25	139.33	22.92	16.45
4.00 & above	152.75	124.45	28.30	22.74
All Farms	154.83	127.96	26.83	21.00

Source: Lavania et. al. (1977), *Economic Affairs*, Vol. 22, p. 354.

It is clear, therefore, that the impact of the green revolution as such on employment is in the direction of greater labour-intensity. The dominant reason for this is the requirements to complete farm operations within a short period of time. This trend of increased employment is particularly useful from the point of view of solving the problems of landless and unskilled labourers, which would not be possible with increased industrialisation alone.

Another favourable effect of the green revolution is the improvement in the income status of landless agricultural workers. This fact may be clear if we look into the trends in real wages of agricultural labourers in the country in the pre-green revolution and post-green revolution period. How far the green revolution has slackened or accentuated the pace of inter-regional variations in agricultural wage rates will be an interesting and important branch of study to understand the economic and social implications of the green revolution. Due to unavailability of data pertaining to inter-district variation in wage rates, an attempt is not made here to analyse this aspect.

On the other hand, HYV agriculture has a number of drawbacks. Some have argued that the rapid agricultural change associated with the green revolution has increased inter-regional and inter-personal inequalities (Nicholson, 1984). Critics have directed their attention towards the problems of the small farmers, who have not been able to participate fully in HYVs cultivation because, as indicated earlier, it is expensive. The adoption of modern inputs and diversification into different crops generally entails a considerable increase in expenses. For instance, it generally takes about 190 kgs. of chemical fertilisers to treat one hectare of HYV paddy in Karnataka, and this costs about Rs. 265. Although HYV seeds are given to the small farmers free of charge by the community development officers, to be productive, farmers still need other inputs such as irrigation equipment — tubewell, pump set, pesticides, fertilisers, and the hiring of more labour all of which involve more expenses. These costs are a barrier to the small farmers for adoption of new farm technology. To counter this barrier, an adequate institutional credit system and the provision of medium and minor irrigation facilities are of utmost importance.

The lessons from past experience in India reveal that the introduction of the "green revolution" in the mid-60s created inter-regional and inter-personal inequalities. But these disparities reflected the fact that governmental investment in irrigation and drainage had concentrated on the already better-endowed regions. Moreover, in only a few areas was the agricultural sector supplied with electric power adequate to support groundwater and surface water development. Notably, efforts to develop Punjab and Haryana as exporters of foodgrain led to heavy investment in irrigation, rural electrification, and the construction of the huge Bhakra and Nangal multi-purpose projects on a perennial river of Himalayan origin. By 1950, Punjab accounted for about half of the cropped area covered by government's irrigation works (Ray et. al., 1979). Similarly, by 1970, per capita electric power consumption was growing at 8.5 per cent and 7.3 per cent annually in Punjab and Haryana, respectively, as compared with 4.2 per cent for India as a whole.

In other regions, the benefits of the green revolution were accumulated by the rich who can afford to invest in the exploitation of surface and groundwater and in chemical fertiliser. Consequently, the disparities between the regions with and without public investment widened. There is no doubt that the green revolution has made a tremendous impact on the development of the national economy. Impressive increases have been registered in the use of tractors, oil engines, pump-sets, tubewells, and in the consumption of fertilisers. Rapid strides have been made in foodgrain production, which reached another plateau in the late-1970s and early-1980s (Table 15.2). In 1987 the Indian Food Corporation has 23.5 million tons of foodgrain in storage compared to none 20 years ago. The World Bank Report (1986) commended India for its success in achieving self-sufficiency in foodgrain production. It stressed, however, the need for the continued and substantial net transfer of capital to the country for the creation of employment and the alleviation of poverty.

Certainly, although the intensive cultivation based on the HYV programme has made agriculture more profitable, this has not solved the problems of rural poverty, unemployment and inter-personal and inter-regional disparities. Some people argue that large increases in agricultural output have further aggravated these problems (Chakrawarti, 1973 and Dantwala, 1986). Why has this happened, despite tremendous progress? Partly because, as mentioned above, growth has been uneven and favoured a few localities. Wheat-growing areas have benefited more than rice-growing areas, and besides wheat and rice, no improved seeds have been evolved for other commercial crops. In most parts of the country the peasants are still practising the traditional subsistence type of agriculture. This is because the spread of the green revolution critically depends upon availability of irrigation facilities. Water is the leading input and in most of the country this can only be assured and controlled through irrigation. The production potential of the HYV programme has not been fully utilised due to the inadequate network of irrigation and since development of such a network is the government's responsibility, the authorities are at fault, for the low level of adoption of the HYV programme, rather than the poor motivation of the farmer.

Although 47 major and 470 medium irrigation reservoirs were under operation (by 1980) and 52 more major on-going schemes are likely to be completed by 1988 in different parts of the country, and even though an ultimate irrigation potential of more than 93.2 million acres has been created, still about 72 per cent of the fertile cropped area is unprotected by irrigation facilities. The irrigation potential of available Indian river water and groundwater has been estimated by the Planning Commission (1976) as follows:

River Water	*Million Acre-feet*
1. Total surface flow of river water is about	1,360
2. Usable river water for irrigation	550
3. Unusable water for irrigation owing to physiographic condition	810

Actual utilisation up to the end of the Fourth Five Year Plan period (1974) was about 200 million acre-feet or 36.3 per cent of the usable river water resources, leaving 350 million acre-feet or 63.6 per cent to run to the sea. This indicates the vastness of nature's gift to India in the form of river water. A huge task is still to be accomplished in harnessing these water resources.

	Million Acres
4. Total area which can ultimately be irrigated by irrigation schemes utilising all the usable river water resources	118

5. Area which can be irrigated by minor irrigation schemes of surface and ground water resources. (Surface-36 mln. Ground-43 mln.) 79

Total 197

Thus against the total cropped area of about 385 million acres in the country, only about 197 million acres can be economically irrigated if the entire available water resources are utilised. The remaining 188 million acres will have to depend on monsoons. Only 94.8 million of the 197 million acres had been brought under irrigation by major, medium and minor schemes up to the end of the Fifth Five Year Plan (Table 15.6). This shows that a large percentage of the fertile arable land is still allowed by the Indian economic planners to depend on the vagaries of the monsoon.

TABLE 15.6: Area under irrigation by different sources (1950—1979)

(in '000' hectares)

Year	Major (Canals)		Minor		Other Sources	Total
	Public	Private	Tanks	Tube Wells		
1950-51	71,58	11,37	36,13	59,78	29,67	2,08,53
1955-56	80,25	13,60	44,23	67,39	22,11	2,27,58
1960-61	91,70	12,00	45,61	72,90	24,40	2,46,61
1964-65	98,59	10,99	42,58	86,53	24,75	2,63,44
1970-71	119,72	9,66	41,12	118,87	22,66	3,11,03
1974-75	126,52	8,60	35,44	142,29	24,22	3,37,09
1975-76	129,04	8,63	39,82	143,63	23,79	3,44,91
1976-77	130,31	8,48	39,10	150,62	22,96	3,51,47
1977-78	137,25	8,44	39,01	156,03	24,80	3,65,59
1978-79	142,70	8,38	39,18	164,10	25,25	3,79,61
Percentage change (1950—79)	99.3	−26.2	8.4	174.5	14.9	82.0

Source: Government of India, Publication Division, New Delhi, *Statistical Abstract of India* — 1982, p. 75, Table 33.

The largest spread of the unirrigated areas is in the western and central areas of the country. This vast region comprises the states of Maharashtra, Madhya Pradesh, Rajasthan, Karnataka, Andhra Pradesh and Orissa. These states together account for 179 million acres, or about two-thirds of the country's total unirrigated areas (Naidu, 1975). There are numerous rivers and rivulets in which, adequate water is available. In particular, the rivers of northern India rising in the Himalayas are snow fed and flow all the year round.

With regard to groundwater, there are certain areas in the country which have a very high potential. Although the presence of groundwater is estimated to be almost everywhere in the country, it is found at varying depths and in varying quantities. The most abundant amounts of groundwater are thonght to be economically available in four main zones. The most important are: the Gondwana sedimentary zone — (Barkar and Godavari river basins of Andhra Pradesh and Orissa); the Cenozoic sedimentary zone of Tamil Nadu, Kerala, and Gujarat; the Cenozoic Fault Basin — the rift zone of the Narmada, Tapti and Purna river valley areas in Madhya Pradesh and Maharashtra; and the Ganga (Gangetic) and Brahmaputra alluvial zone of Punjab, Uttar Pradesh, Bihar, and West Bengal. Despite the great quantities of groundwater available in all these areas, relatively little of it has actually been tapped. Part of the problem is that the costs and type of equipment needed to extract it is beyond the capacity of an average farmer. There is not much governmental effort to develop these areas to their full capacity due to want of funds and other facilities.

Ironically, with regard to the existing major irrigation projects, there has been a low percentage of utilisation of water compared to the potential created. This under-utilisation of the potential has been mainly due to the delay in the construction of field channels and lack of other coordinated development work in the command areas. India cannot afford to be slack in this respect in view of the vast amounts of money which have been invested on these large-scale projects. And in view of the crucial role of water in stepping up the production capacity of the marginal farmers, it is of paramount urgency and importance that the authorities should concentrate on geographically decentralised and dispersed minor irrigation schemes.

Priority to Minor Irrigation

Irrigation is not only a prerequisite for increased agricultural production but the type of irrigation is also important to reduce the regional differences in land productivity. Types differ in their ability to supply water when and where it is needed to the farmers. As is generally known, the monsoons are erratic in nature. After a brief period of uncertainty (sometimes 3 to 4 weeks) flooding rain usually occurs. The lakes, rivers and streams will overflow from the delayed downpours. Therefore, a given amount of water will be helpful to farmers if it can be delivered in those areas where it is most needed at the particular week of the sowing season in June.

In the circumstances, a few large-scale, multi-purpose reservoirs, in a few localities, do not serve the needs of most farmers, whose farmholdings are scattered over a wide geographical area. We suggest that taking advantage of all rivers, perennial streams and groundwater resources, investment in irrigation should be made in a decentralised way to meet the needs of all regions. The Japanese experience indicates that inter-regional equity in land productivity was achieved through direct investment in small-scale irrigation. The statistics with regard to irrigation schemes in India show that the initial Five Year Plans concentrated too heavily on large-scale irrigation projects at the expense of other minor irrigation methods. The First Five Year Plan placed almost exclusive emphasis on large scale projects, and although the emphasis decreased somewhat in the next three plans, nevertheless, in all of them expenditure on large scale projects substantially exceeded that on medium and small-scale ones (John Mellor et. al., 1968).

As Indian economic planners had been initially intended, multi-purpose schemes have not been capable of supplying enough water in the command areas. Although 47 key reservoirs have been constructed at great cost, over time they have been silting up rapidly, due to deforestation and denudation of pasture and other non-agricultural lands which expose the top soil and leave it defenceless. A greater emphasis on afforestation and pasture development should be taken up in the catchment areas. This will not only reduce soil losses but also increase the potential availability of underground water. Otherwise, the reservoirs will lose their capacity to hold water for the droughts which regularly occur every several years.

Another problem is that the dams are used not only for irrigation, but also, primarily, for hydroelectric power generation. This means that a certain minimum level of water must be maintained behind the dam. During the monsoon season, when the demand for water is less acute, a large supply of water is available in the reservoirs, whereas during the dry season, when water is really needed, the dams can serve only a portion of it. Thus, the reservoirs have not been able to meet the full demand for irrigation. For example, although the canal irrigation system in the northern states of India, like Uttar Pradesh, Punjab and Haryana, is generally regarded as the most efficient and most equitable in the country, it has become much less reliable. Available statistics show that increasing numbers of farmers in these areas are adopting tubewell or other minor methods of irrigation since these methods are controlled by an individual farmer. Between 1960-61 and 1979-80, Punjab's net irrigated area increased from 2.02 million hectares to 3.52 million hectares (or from 54 per cent to 78 per cent of net sown area, see Table 15.7), and Haryana's net irrigated area increased from 1.01 million hectares to 2.17 million hectares (or 30 per cent to 61 per cent of net sown area). By 1981-82 there were 610,000 tubewells in Punjab and 343,000 in Haryana. As of 1980, 42 per cent of Punjab's net sown area and 21 per cent of Haryana's were irrigated by tubewells, compared with the all India average of 6 per cent. In terms of coverage, tubewells accounted for 96 per cent of the area irrigated in Punjab as of 1978 and 98 per cent in Haryana as of 1981-82. Similarly, in Uttar Pradesh and Bihar with substantial potential for tubewell development, the percentage of net sown area irrigated by tubewells was 24 per cent and 9 per cent respectively. This shows that minor irrigation methods have been displacing canal irrigation systems, which are becoming less advantageous, since they permit little control over the timely delivery and quantity of water provided to the farmers. As noted above HYV crops are highly responsive to fertiliser use, but farmers are likely to apply fertilisers at higher levels only if they are sure about the timeliness and quantity of water available for the crop. Since minor irrigation methods allow the farmers much greater control over water application, they are more suitable for High Yielding Variety cultivation.

In view of the above problems, and to avoid the huge investment in constructing large-scale projects, priorities should be shifted to small and medium-scale irrigation works. First of all, systematic surveys should be conducted to explore the possibilities of using local water resources to devise innovative methods of irrigation which are appropriate to specific local conditions. The necessary research to determine the relative costs and benefits of an alternative system of irrigation should be undertaken. Few such studies have been carried out by the Indian Planning Commission.

TABLE 15.7: **Area irrigated by different methods of irrigation in Punjab & Haryana (1960-1982)**

Methods	1960-61	1970-71	1981-82	Percentage Change 1960-1982
Punjab :				
Canals	1,173	1,286	1,415	20
Wells	829	1,591	1,976	138
Others	18	11	17	−5
Total	2,020	2,888	3,408	69
Net sown area irrigated (%)	54	71	81	
Haryana :				
Canals	824	965	1,183	44
Wells	166	594	1,034	523
Others	17	6	16	−40
Total	1,007	1,565	2,248	123
Net sown area irrigated (%)	30	44	61	

Source: Government of Punjab, *Statistical Abstract of Punjab*, 1983, Chandigarh, p. 197, Table 8.1.
Government of Haryana, *Statistical Abstract of Haryana*, 1982-83, Economic and Statistical Organisation, Chandigarh, p. 141, Table 7.1.

Field observations have shown the benefit/cost ratio of minor irrigation schemes to be greater than that of large-scale projects (Reidinger, 1980). Moreover, they can be completed quickly and thus are immediate and reliable sources of irrigation in the hands of an individual farmer. The minor irrigation programme is essentially implemented through individual efforts, and should be supported by adequate financial assistance through institutional sources. The pump-sets, flexible pipes, steel pipes and other equipment necessary for shallow tubewell and lift irrigation schemes are, in cost, beyond the capacity of marginal, poor farmers. For example, to buy and install a shallow tubewell generally costs Rs. 5,600 to 6,800 and the maintenance and running costs are in addition, about Rs. 800 to 900 a year. Therefore, it requires an adequate credit system. But the bureaucratic administrative procedures typically involved need to be simplified.

A few medium-scale public lift irrigation projects have been constructed along the course of the river Tungabhadra in Karnataka and they are operating most successfully. Such medium-scale river pump systems are usually capable of irrigating three or four villages near the river. This method has become a most appropriate one for drought affected areas of Karnataka. Another system, the deep tubewell, is one where the well is sunk much deeper — varying from 200 to 800 feet. A deep tubewell can irrigate around 200 to 250 acres of land. Needless to say, the pump in both these cases are expensive (about Rs. 50,000) and by and large have to be established and run by the government.

The minor irrigation programme up to the end of Fourth Five Year Plan suffered on account of the slow rate of rural electrification and insufficient flow of institutional credit for shallow tubewells and the purchase of pumpsets. Table 15.8 shows the limited amount of electric energy supplied to irrigation. It is evident that of the total energy produced, only 7.3 per cent is made available to irrigation, whereas 70.5 per cent is sold to industries. One can conclude from this that the appropriate infrastructure for agricultural production had not been built up to the end of fourth Five Year Plan. This is one measure of the government's failure to create equity and employment within the rural and agricultural sector.

TABLE 15.8: Electric power sold to industry and irrigation (1967-68)

(in million kWh.)

States	Total Generation	Energy Sales to the States	Sales to Industries	Sales to Irrigation
Andhra Pradesh	1,589.3	1,034.9	576.8	172.6
Assam	97.5	59.2	22.5	0.1
Bihar	3,231.9	2,021.4	1,374.9	29.3
Gujarat	2,207.2	1,796.4	1,421.1	100.5
Jammu & Kashmir	93.6	102.5	40.1	1.9
Kerala	841.9	749.8	602.9	24.1
Madhya Pradesh	1,206.9	1,028.4	859.2	11.7
Tamil Nadu	4,209.1	3,216.9	1,793.6	817.6
Maharashtra	5,650.8	4,760.6	3,257.7	89.5
Karnataka	1,875.3	1,381.6	1,035.5	66.1
Orissa	947.8	976.6	890.9	1.7
Punjab	3,415.6	2,306.3	1,906.1	213.5
Rajasthan	164.9	325.7	197.8	19.5
Uttar Pradesh	3,030.3	2,371.1	1,639.1	315.8
West Bengal	3,874.9	4,099.8	3,018.5	2.7
Others*	414.7	683.8	217.4	25.6
India	32,851.1	26,915.4	18,854.8	1,892.2

* Others include all Union Territories & Minor States.

Source: Government of India, Administrative Reform Commission, *Report of the Study Team*, Vol. II, Annexure XIII (A), p. 785.

REFERENCES

Ahluwalia S. Montek, 1977 : "Rural Poverty and Agricultural Performance in India", *Journal of Development Studies*, Vol. 13 (July, 1977), pp. 298-310, Table 3(b).

Annual reference, *India* 1984, Government of India, Publication Division, New Delhi.

Bruce F., Johnston and William Clarke, (Ed.), 1982 : *Redesigning Rural Development*, Johns Hopkins University Press, Baltimore, p. 72.

Chinnappa, N., 1977 : "Rice growing areas in Tamil Nadu", (in) Farmer, B.H., (Ed.), *Green Revolution? Technology and Change in Rice Growing Areas of Tamil Nadu and Sri Lanka*, Westview Press, Boulder, Colo., Chapter 8, p. 215.

Dantwala, M. L., 1986 : "Technology Growth and Equity in Agriculture", (in) John Mellor (Ed.), *Agricultural Change and Rural Poverty*, The Johns Hopkins University Press, Baltimore and London, pp. 110-15.

Friedmann John, 1974 : *A Spatial Framework for Rural Development: Problems of Organisation and Implementation*, University of California, Los Angeles, p. 4.

Government of India, Planning Commission, New Delhi, *Percentage of Population Below the Poverty Line by States*, New Delhi, p. 135.

———, *Fifth Five Year Plan*, document (1974-79), Chapter 8, pp. 87-93.

———, 1979 : *Report of the Study Team*, Vol. II, p. 136.

———, *Sixth Five Year Plan*, document (1980-85), pp. 17-18.

Government of India, Publication Division, New Delhi, Annual Reference, *India 1984*, pp. 213-16.

———, Annual Reference, *India 1984*.

Government of Punjab, 1983 : *Statistical Abstract of Punjab*, pp. 194-195.

High Commission of India, Ottawa, *India News*, Weekly News Bulletin, February, 1987, p. 1.

India Abroad, Weekly News Paper, Fort Erie, Ontario, August 14, 1987, p. 6.

Iqbal, B. A., and Mohammad, T., 1986 : "Sustained Agricultural Production", *Food Policy*, Vol. II, No. 2, pp. 94-98.

Krishna, R., 1986 : "Economic Outlook for India", (in) Tames Koach, (Ed.), *India 2000: The Next Fifteen Years*, Allied Publishers Pvt. Ltd., New Delhi, pp. 170-178.

Lavania, G. S., Bhalerao, M. and Jain, B. L., 1977 : "Modern technology and human labour utilisation in developing countries — A sample study in India", *Economic Affairs*, Vol. 22, pp. 352-58.

Micheal Lipton, 1977 : *Why Poor People Stay Poor*, Harvard University Press, Cambridge, pp. 24-28.

Misra, R. P., and Nataraj, 1981 : "India: Blending Central and Grassroots Planning", (in) Stohr and Taylor (Eds.), *Development From Above or Below?*, John Wiley and Sons Ltd., pp. 259-79.

Naidu, I. J., 1975 : All India Report on Agricultural Censes (1970-71), Ministry of Agriculture and Irrigation Publication, New Delhi, p. 57.

Narain Dharam, 1965 : *The Impact of Price Movements on Areas Under Selected Crops in India*, Cambridge University Press, pp. 43-46.

———, 1976 : "Growth of Productivity in Indian Agriculture", *Journal of Agricultural Economics*, Vol. 32 (Jan-Mar. 1977).

Reidinger, R. B., 1980 : "Water Management by Administrative Procedures in an Indian Irrigation System", (in) E. Walter Coward (Ed.), *Irrigation in Agricultural Development in Asia*, Cornell University Press, Ithaca, p. 265.

Swindale, L. D., 1981 : Coromandel Lecture on "A time for Rainfed Agriculture", 10th, December 1981, (in) *Times of India*, New Delhi, p. 1 and 4.

The Economist, Newsweekly, May 9, 1987 : "India Flying Solo", pp. 14-15.

World Development Report (1982), Oxford University Press, p. 61.

16

SOCIAL ASPECTS OF RURAL DEVELOPMENT IN INDIA

SHAM SHER SINGH AND C. R. PATHAK

RURALITY IS the spinal cord of socio-economic, political and cultural life of India, but the saddest story of its being in the grips of centuries-old poverty is not unknown to anybody. Though many efforts have been made to improve the conditions of rurality, yet the sky remains overcast with only a few peeping blue patches of development. Rather, it has been proved that the rural poverty is increasing both in terms of percentage households below the poverty line and in the intensity of poverty (Rao, 1979). Failure of previous programmes like Community Development Programme is attributed to many factors such as limited resources, failure to fully exploit local resources for increasing the national income, sectoral approach in planning, incompetent implementing agencies, urban bias in development and much neglected socio economic structure and changes therein (Saini et al., 1977). Many of these factors have been emphasised and explained by a number of writers but the factor of the negligence of social aspects of rural people, which has great bearing on the fruitful implementation and consequently development of the villages, still remains unexplored by the planners though it has been given thought off and on. It is the purpose of this paper to throw light on the societal aspects like caste, class and factions and their impact on the development of the villages. Some solutions are also put forward to neutralise the negative impacts.

The Village India: Myth and Reality

A village is often considered as a community — a cohesive social group unified by the harmoneous relationship of individuals, family, kinship and caste; and working in a democratic way to provide solidarity to the village society itself. But the reality is something else. Our villages are suffering from discontent, violence against person and property, hunger, disease, illiteracy, ignorance and stupidity, free fights for status, land and water, caste rivalries, group conflicts, corruption and bribery, social and individual disorganisation, social tensions, absence of civic amenities, chronic under employment and sub-human standard of living, economic dependence and even helplessness, false dignity of labour, immitation of city ways, caste distinction and segregation, untouchability, uncleanliness, inconsistency, and disregard of any moral code of ethics characterises the rural solidarity, cohesiveness, mutual consideration and cooperation or simplicity of living and adherence to accepted norms (Baljit Singh, 1961). These realities are mainly due to different socio-cultural and economic groups coupled with their psychological implications.

The village life is encompassed by two systems in terms of hierarchical principles. First is the caste hierarchy and the second the economic class hierarchy which has become more intense during the past few years. These two systems act and react on each other and also overlap and consequently give rise to many factions which are generally motivated by self-interest and "cut across boundaries of structural units" (Dubey, 1975) of the rural society. These structural units are different caste and class groups. This factionalism in Indian rural society is not new, rather, the village for long has been a faction society. "Factional splits are the reality and the active forces and solidarity and cohesiveness of the group have been a myth like several other characteristics" (Baljit Singh, 1961). And as Karve (1954) points out these factional splits should be taken fully into account in any programme of rural development. Therefore, the origin, nature and process of factionalism should be fully understood.

The Factions

Principally, the village India is divided into hierarchical groups of different castes which are "closed to each other and shut all doors to social mobility" (Baljit Singh, 1961). This hierarchical caste system is followed by the economic class hierarchy following almost the same order as generally higher castes are economically better than the lower castes. Higher castes are the landowning class and the lower caste, the landless class, creating two major economic classes in the rural society. Of course, marginal farmers and farmers having large tracts of land create two more classes. Thus, the whole rurality is divided into four broad economic classes.

These castes and classes being on the same track make the village society self-contained in social affairs but inter-dependent in economic affairs and prepare ground for another devisive but decisive force, i.e. the faction. Factions are social alignments on the basis of self-interest, and intersect through caste and class groups. "These are an old, ingrained pattern in village life and must be considered as a basic structural aspect of traditional village organisation along with caste, 'tolas', 'pannas', 'gotras' and other groupings" (Lewis, 1953).

As hinted above, the factions develop through schism among villagers to fulfil certain interests. These interests might be of positive or negative nature and are generally related with land, social status and power, ego satisfaction and ideological differences. Two forces of : (i) external environment like abolition of zamindari by law and (ii) the internal elements like individuals work as catalysts in the formation of factions (Rastogi, 1975). Formation generally follows caste or class lines or kinship lineages. Members of a faction cluster around a focus, i.e., the leader who may be a dominating person or an elderly one in case of kinship lineage. A faction may be ephemeral which comes "up over a particular issue in conflict and die" out as soon as the conflict is resolved, or they may have stability over time and may represent an aggregation of interests involving a continuous tussle (Dubey, 1975). The size of a faction changes as the group situation changes in response to external and internal factors affecting the interests and values of the individuals' (Rastogi, 1975).

Factions in a village not only "involve the entire village divided and fighting but their litigation and fights in courts and outside involve a number of other villages. Due to the practice of caste endogamy and village and Gotra exogamy each household has patrilineal and matrilineal relations in other villages" (Baljit Singh, 1961). Thus, a village may be connected to as many villages as the number of households in it through marriages, etc. Relatives from other villages come to support their relatives whenever needed and, since they have their own factions in their villages, factional dispute in one village involves a number of villages resulting in the formation of inter-village factions. This involvement of many villages also takes place when a village caste panchayat fails to satisfy a faction and the aggrieved party calls for a bigger caste panchayat.

The study of rural faction society has recently lured the attention of sociologists but "the kinship rivalries for familial supremacy and status" in the "form of factions along caste lines have been tearing as under the rural communities since ages" (Baljit Singh, 1961). Cohn (1954) in his study of Madhopur (Jaunpur district) points out that it did exist in the 19th century and there were fights, killings and house burnings usually involving land ownership and arguments over land. But the factionalism has increased "since the onset of steps like zamindari abolition, adult franchise and panchayati raj" (Rastogi, 1975), "reflecting the growing impact of urban contracts, pressure of democratisation and modernisation besides increasing political consciousness" (Rastogi, 1975). Recent infiltration of politics particularly into the Panchayati Raj has intensified factionalism. Although it cuts across the social and economic barriers, still a distinct caste and class realignment has been taking place and creates its own vested interests.

Rural Development: Its Nature and Processes: A Theoretical Framework

As Brookfield (1975) suggests, 'The popular trend is to define development in terms of progress towards a complex of welfare goals, such as reduction of poverty and unemployment, and diminution of inequality". The ingredients of this definition are digestible by the concept of rural development also.

The beginning of the study of development processes in the sociological context is attributed to such writers as Spencer, Tylor, Morgan, Durkheim and Marx (Norman Long, 1977). Long puts all the writers into two broad camps which seem quite convincing. In the first camp come those who support a neo-evolutionary modernisation approach, and in the second camp those who espouse a Marxist orientation. Modernisation approach is fed by three processes of differentiation, integration and adaptation. Smelser (1963) believes that when society starts developing through the processes of modernisation of technology, commercialisation of agriculture, industrialisation and urbanisation, as discussed by Lewis (1953), the structural differentiation occurs as more specialised and autonomous social units are formed. Integration takes place when those differentiated structures are united on a new basis, e.g. cooperative enterprises, etc. (Norman

Long, 1979). And the third process emphasises that the developed or developing society possesses an institutional structure which can absorb changes beyond its own initial premises and deal with structural dislocations, protest movements and elements of disorganisation (Eisenstadt, 1966). The capability of absorption, of course, will vary from one social system to another.

The Marxist approach of development concentrates itself on the emergence of capitalism, exploitation of labour and then revolutionary transformation from capitalist to socialist society, where the "dictatorship of proletariat" is established (Jhingan, 1969).

Coming to the development processes in the case of rural India, the first approach, i.e. neo-evolutionary modernisation approach, finds some imprints but it is very difficult to find out which process is working at which stage because Indian society is formed of many stratifications like caste, class, factions, etc., which cut across economic, socio-cultural and institutional stratifications. Even then a mixed impact of these processes of differentiation, integration and adaptation is visible as the rurality is developing as well as changing. For example with the advent of urbanisation and other processes of development the caste bonds are becoming loose, modern thought and technology have started stepping in and new class of rural elites is emerging which is educated and more open to urban influences, and sometimes politically energised.

Rural Development Programmes in India: A Critique

In India major attention to villages had been given only after independence. Britishers tempered the traditional economy and polity of the villages to fulfil the exploitative objectives. After independence two major policy decisions were taken to stimulate the growth in the rural sector. The first was to effectively deal with the problem of landlordism and the second was to start the Community Development Programme. This programme had two broad objectives: "to provide for a substantial increase in the country's production, and for improvements in the system of communication in rural health and hygiene, and in village education" and "to initiate and direct a process of integrated cultural change aimed at transforming the social and economic life of the villages" (Dubey, 1958).

In practice this programme of rural development was limited mainly to agricultural extension activities and there too "about 70 per cent of the benefits went to the affluent farmers — poorer small-holders were often not in a position to take full advantage of the new methods" (Norman Long, 1977). On the other hand needs of artisans and agricultural labourers were ignored. The net result of this programme is that large farmers have become wealthier and the poor strata still remain neglected adding another conflict of poorer and wealthier villagers to already existing faction-ridden society; and it has also been found by Dubey (1958), while analysing the effects of the programme in Uttar Pradesh, that the adoption of certain improved agricultural implements such as mechanical ploughs and tractors led to a reduction in the employment opportunities for labourers in the area. It has also been observed that acceptability of the projects initiated under the programme was effected by the economic advantage and utility, the kind and prestige of power that the project might offer to the social group concerned, curiosity value, e.g. a novel thing is adopted but abandoned later if it does not offer any practical benefit, and the question of compliance with the wishes of the Government or influential village leaders. The poorer section was found to behave in a subservient way to the demands of the Government indicating the bureaucratic way of implementing the programme which generally does not bother about the social values.

Village Society and the Development Problems

Having known the nature of village society and rural development in theory and practice, now the real issue is, after all where do the sociological aspects inhibit the development process and why many programmes go away? Following are the important issues involved in the process.

The co-existence of class and caste hierarchy presents a mix of two conflicting cultures — one is the traditional folk culture characterised by "small, isolated, non-literate and homogeneous" (Redfield, 1947) society with a strong sense of caste and group solidarity where "members act with reference to each other by tacit and traditional understanding, no formal contracts or agreements are needed and custom fixes everything" (Redfield 1947); and the other is so-called modern culture or urban culture where traditional caste bonds give way and the show of economic status dominates the scene. This mix of culture of old and new ideas has created many differences, integrations and adaptations. For example the zamindari abolition, a socialistic step, could not be digested by the village society it has left a void, a sort of power vacuum that has hastened the process of social disorganisation and given strength to the forces of conflict and antagonism" (Baljit Singh, 1961). The class and caste-ridden society was not prepared for this change. And introduction of the Panchayat System added fuel to the fire. Right from elections to the decisions of Panchayats, the caste and class play their dirty game in the system. In a village, in district Mohindergarh, Haryana, a well-educated retired military person could not become "Surpunch" because he was a scheduled caste member (Singh, 1979).

The very structure of the Indian caste system presents a hierarchy of occupation also, e.g. knowledge is meant for Brahmins, defence for Rajputs, productive activities for Vaishyas and services for the Sudras. Hindu society is still trying to stick to this system resisting any changes especially in the case of Sudras. Even in the intelligentsia and the educated section a "Sudra" is looked down upon. What development will do in this sort of fractured society where the loyalty of an individual is not towards the community but towards the caste? "Hence the caste relations have become anachronistic and the system as a whole the impediment to any progress" (Baljit Singh, 1961 : 8). Moreover casteism is being used for vested political and economic interests. "The village factions" which are explained later in detail, "support and feed it" even though there is law to the contrary, (Baljit Singh, 1961). For instance, untouchability has been abolished by law but it is still being practised in the villages and the depressed and the weak find themselves helpless against powerful factions of the landed upper peasantry.

Coming to the impact of class hierarchy, it is observed that the lower landless class is generally deprived of whatever benefit is acquired from any programme. Even financial help is given on some guarantee which all landless poor people cannot give. For this they have to run after some landlord who makes use of the opportunity to suck the blood of the poor in the form of his so-called "commission". Among farmers the benefits are converging towards affluent farmers. It has been proved that in 1950-51, farmers having land up to five acres used to market 33.6 per cent of their output, but in 1960-61 it came down to 18.77 per cent, whereas the rich peasantry having land 50 acres and above used to market 51.4 per cent of their produce but it went up to 61.87 per cent in 1960-61 (Sau, 1973). This also indicates that in rural areas capital is being concentrated in the hands of "a few" and the gap between the rich and the poor is widening. Sau (1973) is also of the view that the "capitalism has penetrated Indian agriculture". Cooperatives etc. . .meant for the poor farmers are also not free from the illinfluence of landlords and the rich peasantry.

"The phenomenon of bigger farmers reaping the economic benefits of improved technology has recently been interpreted as a consequence of urban bias of Indian planning" (Lipton, 1968). This is explained by him by counting the factors such as : (a) marketed surplus is primarily needed for providing resource for industrialisation; (b) the neglect of village power structure which funnels services and capital to the people of higher politico-economic status; (c) the close connection between rich farmers and money-lenders who usually have property in the town and wish to protect their urban resources and power; (d) the planners themselves are pressurised by these and other groups; and (e) the urban orientation of extension workers and other concerned with rural development. Another point which Lipton has missed is that there landlords work as "control points" for the administrative and political interests in rural areas and in return they get such facilities as loans, etc. . .which make them economically and politically stronger. Hence, all these programmes aiming mainly at agricultural development have brought the capitalistic trend in Indian rural economy and have failed to bring any qualitative change in the socio-economic and power structure of the rural areas.

The major hindrance in the way of development of villages is the "faction society" which is characterised by such features as division, litigation, feuds, quarrels and fights, crimes and acts of violence, etc. . . . Solidarity of an individual faction is used "to oppose and pull down other rather than to promote common welfare. A faction narrows down the area of communication and social intercourse within a caste or society and canalises the energies and resources of various lineages and kinship groupings within it against each other" (Baljit Singh, 1961). Obviously, under these conditions society cannot respond to any progress and its entire dynamism is engulfed by the efforts of factions. Its overall ordering law becomes regressive and it makes the village static and hostile to change inspite of their poverty, needs and aspirations. The situation is further aggravated by the factional relations among villages which help in maintaining the factional disorganisations as a continuous process.

Now, these two aspects of society, viz. factional pattern and hierarchical organisation, make the village society even more complex than the urban society. It enables feudalism and caste system survive and grow in spite of legislations. Moreover, the external factors affecting the faction society make progress extremely difficult. They react sharply to any change with a mere suspicion that the change may place their opponent in a more favourable situation. Therefore, fights for the status quo are always respected so that the opposition faction may not gain or improve. Progress itself becomes unwanted because while it may benefit them it may also benefit their opponents and they would like to prevent it at any cost or sacrifice. Thus, undoubtedly, the faction society hinders all progress and also explains why village people in general are so indifferent to the plans and the development programmes.

Some Guidelines

There are two different approaches for the social change: the "improvement approach" which believes in "progressive movement" of society without any radical changes in the traditional and legal system (World Bank, 1960); and the "transformation approach" which attempts to establish new forms of agricultural and social organisation, and which makes a radical breakthrough in the existing peasant systems in terms of scale of operation, production techniques and social-legal structure (Norman Long, 1977). The first approach allows for the continuity of existing social institutions whilst the second approach aims at large-scale government intervention. During her planned period India has mainly relied upon the first approach but the situation has worsened as proved by Dubey (1958), Ranjit Sau (1973) and Epstein who studied one village in south India first in 1962 and then in 1973. India did take help of the second approach, e.g. land consolidation and the zamindari abolition but the results were not very favourable.

Then, after all, what should be done with the Indian rurality? Following are some major considerations which might be considered in developing and adjusting the changes in the rural society and curtailing the negative features. These considerations are based on a compromise of mixed strategy, as has been supported by Feldman (1969), too. The considerations are:

Traditionally in Indian society there is concentration of economic power in the hands of the landowning class, especially big farmers, which is leading the rural economy towards "capitalistic" and "monopolistic" ways. To abate this converging trend is the first thing which must be done if rural society is to be developed smoothly. A reduction in the economic disparity will first be followed by certain differences, as are happening today (Epstein, 1973), but integration will also accompany the development and then the sense of adaptability will be infused in society. This feature is also being observed in the Indian rural society but the process is at a very low ebb, e.g. adaptation of the right to own land by the untouchables in the Hindu society. Thus, the reduction of economic disparity will automatically bring changes, dynamism will enter society and the "will" to develop will be infused among the ruralities. Of course, to bring these changes the "transformation approach" has to play the dominant role. At least the evil of caste conflict cannot be abolished without this approach.

Another major aspect to be considered which affect the development process is the institutional changes, which are very slow to come about but bring drastic changes in the societal structure. Work of Christian missionaries in the tribal areas of different parts of the world is not hidden from anybody. In the Community Development Programme it was envisaged that existing institutions and agencies such as village councils (Panchayats), cooperative societies, schools and other voluntary associations concerned with rural development should be incorporated into the scheme so that there could be the widest possible representation of people. This approach towards community improvement was not basically wrong but failed due to the dominance of village elites, inter-village politics (Dubey, 1958) and more control by the administrative structure which itself was not competent to deal with such schemes. The success of institutional changes lies only in the dissolution of the faction situation and group and caste rivalries. Of course, the first point mentioned above will be necessary in bringing about this change, but is not sufficient. There are two ways to do so — changing the environment by infusing new thoughts, ideas and ways of living which only an exhaustive education system can bring; and another is a sort of slow poisoning to the situation, i.e. legitimisation of opposition and dissent. First, this will allow the factional situation to go to the extreme and then they will make them realise their mistake and will try to adjust with the situation. This cyclic trend is also being felt in Indian society. Resistance of higher castes, in general, has come down to a great extent. This second way will also help in understanding the nature, wants, needs and aspirations of different groups which the Indian planning lacks.

Conclusion

One of the main reasons of net coming up to the rural development programmes expectations is not understanding the social conditions properly which are intermingled with the existing economic conditions of the Indian village society. Village society is a mixture of different castes, classes and factions which react and interact on and through each other and not only make the society static and full of conflicts but hostile towards changes also. For development it needs re-structuring which has to be done within certain limits of economic, social and political factors and the national goals. The indication here is towards making society more dynamic through economic and institutional changes. These changes, of course, involve longterm planning processes but should not be left out of sight.

REFERENCES

Brookfield, H., 1975 : *Interdependent Development*, London, Methuen & Co. Ltd.

Cohn, 1954 : Quoted by Oscar Lewis in his *Group Dynamics in North Indian Village*, Planning Commission, New Delhi.

Dubey, S. C., 1958 : *India's Changing Village: Human Factors in Community Development*, London, Routledge and Kegen Paul Ltd.

————, 1975 : Foreword to P.N. Rastogi's *The Nature and Dynamics of Factional Conflict*, Indian Institute of Advanced Studies, Simla.

Eisenstadt, S. N., 1966 : *Modernisation: Protest and Change*, Englewood Cliffs, New Jersy, Prentice Hall Inc.

Epstein, T. S., 1962 : *Economic Development and Social Change in South India*, Manchester University Press.

————, 1973 : *South India: Yesterday, Today and Tomorrow*, London, MacMillan.

Feldman, D., 1969 : "The Economics of Ideology", in C. Leys (ed), *Politics and Change in Developing Countries*, Cambridge, Cambridge University Press.

Jhingan, M. L., 1969 : *The Economic of Development and Planning*, P. Ram & Sons, A-18, Rana Pratap Bagh, Delhi-7.

Karve, D. G., 1954 : *A Foreword to Oscar Lewis Group Dynamics in North Indian Village*, Planning Commission, Delhi.

Lewis, O., 1953 : "Peasant culture in Indian and Mexico", Contribution to the book *Village India* edited by MacKim Marriot.

————,, 1953 : "Tepotztan Re-studied: A critique of the Folk-urban Conceptualisation of Social Change", *Rural Sociology*, Vol. 18, No. 2, pp. 121-34.

Lipton, M., 1968a : "Strategy for Agriculture: Urban bias and Rural Planning", in P. Streeten and M. Lipton's, *The Crisis of Indian Planning: Economic Planning in 1960s*, London, Oxford University Press.

Long, Norman, 1977 : *An Introduction to the Sociology of Rural Development*, Tavistock Publications Ltd., 11 New Fetter Lane, London.

Rao, V. K. R. V., 1979 : "Has Rural Poverty Increased in India?" *Eastern Economist*, Vol. 72, No. 12, pp. 627-34 (March 23, 1979).

Rastogi, P. N., 1975 : *The Nature and Dynamics of Factional Conflict*, The MacMillan Co. of India Ltd., Delhi, Bombay, Calcutta and Madras.

Redfield, R., 1947 : "The Folk Society", *American Journal of Sociology*, Vol. 52 (January issue), pp. 292-308.

Saini, N. S., Suri, K. B. and Ansari, J. N., 1977 : "Rural Planning and Development in India", published in *Readings in Urban and Regional Planning in India*, edited by G. B. K. Rao and N. S. Saini, School of Planning and Architecture, New Delhi, pp. 218-234.

Sau, Ranjit K., 1973 : *Indian Economic Growth: Constraints and Prospects*, Orient longman Ltd., 17, Chittaranjan Avenue, Calcutta.

Singh, Baljit, 1961 : *Next Step in Village India: A Study of Land Reforms and Group Dynamics*, Asia Publishing House, Bombay.

Singh, Sham Sher, 1979 : *An Approach to Village Planning — case study from district Mohindergarh, Southern Haryana*, an unpublished M.R.P. thesis submitted in the Department of Architecture and Regional Planning, I.I.T., Kharagpur, West Bengal.

Smelser, Neil, 1963 : "Mechanism of Change and Adjustment of Change". in B.F. Hoselitz and W.E. Moore (eds.), *Industrialisation and Society*. The Hague: Mouton in collaboration with UNESCO. Reprinted in G. Dalton (ed.), 1971, *Economic Development and Social Change*, New York, The Natural History Press.

INDEX